Animal Breeding
Biotechnology

Animal Breeding Biotechnology

David Thompson
Editor

KOROS PRESS LIMITED
London, UK

Animal Breeding Biotechnology

© 2012

Printed in 2017 for Sale in the Indian Subcontinent

Published by
Koros Press Limited
3 The Pines, Rubery B45 9FF, Rednal,
Birmingham, United Kingdom

Tel.: +44-7826-930152
Email: info@korospress.com
www.korospress.com

ISBN: 978-1-78163-016-7

Editor: David Thompson

British Library Cataloguing in Publication Data
A CIP record for this book is available from the British Library

10 9 8 7 6 5 4 3 2 1

Exclusively distributed by CBS Publishers & Distributors Pvt. Ltd.
Sales & Distribution Rights only for India, Pakistan, Bangladesh, Sri Lanka, Nepal and Bhutan.This book is not to be sold outside these territories.

Contents

Preface

The first decade of the 21st century has been a golden time for the advancement of genomics, driven by the completion of the Human Genome Project (HGP). Various methodologies and technologies have been developed during and after the process of building the human genetic blueprint that have been directly transferred into the studies of domestic animal genomics. The search for genetic underpinnings of human diseases perplexed researchers for many years. Only recently did the genetic factors underlying various human diseases begin to be revealed, especially with the help of genome-wide association studies (GWAS) using SNP arrays. Single nucleotide polymorphisms (SNPs) are bi-allelic genetic markers, and they are easy to evaluate and interpret and are widely distributed within genomes. With proper coverage and density over the whole-genome, SNPs could capture the linkage disequilibrium (LD) information embedded in the genome, which could be used to pinpoint genes underlying human diseases. For domestic animals, these tools can contribute to i) better understanding of species' evolution, domestication and breed formation, and developing new theories of population genetics; ii) dissecting the genetic mechanisms of complex agricultural traits; and iii) improving selection methods for genetic improvement of animal production.

High-density SNP arrays were built for important farm animals, first for those with reference genomes and then recently also for others without reference genomes with the advent and application of the massive parallel sequencing technologies. The preparation and utilization of SNP arrays are having considerable impacts on the theory and practice of animal breeding and genetics, which will play important roles in the years to come. In this review, the whole-genome sequencing and HapMap studies of several important domestic animals are briefly summarized. Then, details about the development of SNP arrays and their applications in various genetics and genomics research projects are also reviewed. Lastly, lessons learned from the reported studies and prospects for future work are discussed.

The whole-genome sequencing strategies for most domestic animals were taken directly from human genome sequencing, i.e., combining both whole-genome shotgun (WGS) and BAC-to-BAC sequencing. Based on their significance in agriculture and as biomedical models, chickens, dogs, cattle, horses and pigs have had their genomes sequenced, as well as some other important animals. Due to the rapid development of "next-generation" sequencing technologies and the availability of reference genomes, these strategies have been modified for different species. For those well-studied species in which high levels of genomics knowledge and sequence coverage are required, such as chicken, dog, cattle, horse and pig, the dual approaches of WGS and BAC sequencing were applied.

This book places this subject in its most modern context as the foundation of this subject It is an authoritative resource of the subject that will benefit researchers, students, and teachers alike.

—Editor

1

Climate Change and the Characterization, Breeding and Conservation of Animal Genetic Resources

Animal genetic diversity is critical for food security and rural development. It allows farmers to select stocks or develop new breeds in response to changing conditions, including climate change, new or resurgent disease threats, new knowledge of human nutritional requirements, and changing market conditions or changing societal needs – all of which are largely unpredictable. What *is* predictable is increased future human demand for food. The effects will be most acute in developing countries, where the increase in demand is expected to be greatest, and occur at a rate faster than increases in production, and where climate change is projected to have its greatest impact.

The global livestock sector is characterized by a growing dichotomy between livestock kept by large numbers of smallholders and pastoralists in support of livelihoods and rural food security, and those kept in intensive commercial production systems. FAO's latest global assessment of breed diversity identifies 7040 local breeds (each reported by only one country) and 1051 transboundary breeds (each reported by several countries) (FAO, 2009a). A breed is a cultural rather than a biological or technical entity (Eding, 2008). A breed covers groups of animals having similar characteristics that depend on geographical area and origin. Most European breeds are well defined, distinct and for a large part genetically isolated. In contrast, Asian and African breeds most often correspond to local populations that differ only gradually. About two-thirds of reported breeds are currently found in developing countries. Local breeds are commonly used in grassland-based pastoral and small-scale mixed crop-livestock systems where they deliver a wide range of products and services to the local community, with low to medium use of external inputs.

They are usually not well characterized and described, and seldom subject to structured breeding programmes to improve performance. So-called "international transboundary breeds" of the five major livestock species (cattle, sheep, goats, pigs, chickens), many of them high-output commercial breeds, have spread globally for use in large-scale, often landless, production systems, where they produce single products for the market (either milk, meat or eggs) with high levels of external inputs. Most flows of genetic material occur among developed countries, most of which are without zoosanitary restrictions, and involve animals suited to high-input production systems. More than 90% of exports originate from developed countries, and the share of trade in genetic material from developed to developing countries increased from 20% in 1995 to 30% in 2005 (Gollin *et al.*, 2008). In many cases, the improved components of the high-input management systems needed to express the genetic potential of the high-output breeds have been transferred to developing countries. Industrial systems utilizing sophisticated technology and based on internationally sourced feed and animal genetics already produce 55% of pork, 68% of eggs and 74% of poultry meat globally.

For 36% of breeds the risk status is unknown (FAO, 2009a). The loss of within-breed diversity is not known, although within commercial breeds, high selection pressure, particularly when combined with poor breeding programmes, leads to a narrowing genetic base. About 9% of reported breeds are extinct and 20% are currently classified as being at risk. The species involved in production and marketing systems with fast structural change show high proportions of breeds currently at-risk and already extinct. This includes 38% of chicken breeds, 35% of pig, 33% of horse, and 31% of cattle breeds (FAO, 2009a). Economic and market drivers also made up 28.5% of all responses in FAO's questionnaire survey on threats to animal genetic resources across the main species (FAO, 2009c). It can be expected that multifunctional local breeds continue to play a role in the livelihoods of poor people and in marginal areas.

Livestock production contributes to and will be affected by climate change. FAO has promoted discussion of the environmental impact of different livestock production systems. Consideration of the livestock sector is crucial for adaptation to, and mitigation of, climate change – because the sector is a large producer of greenhouse gases (GHG). Eighteen percent of global GHG emissions are attributed to livestock – via land use and land-use change (directly for grazing or indirectly through production of feed crops), manure management, and enteric fermentation (FAO, 2006a; FAO, 2010).

Climate change will affect the products and services provided by agricultural biodiversity. But this biodiversity has not yet been properly integrated into strategies for adaptation to and mitigation of climate change. Its role in the resilience of food systems still needs to be addressed. The Intergovernmental Panel on Climate Change (IPCC) report on biological diversity (Gitay *et al.*, 2002) and likewise the report of the Convention on Biological Diversity on climate change (CBD, 2009) contain little mention of agricultural biodiversity, and a recent literature review (Campbell *et al.*, 2009) largely ignores livestock diversity. In a survey on threats to livestock diversity (FAO, 2009c), climate change was only mentioned as a minor factor in the context of extensive land-based production systems. The findings show that many stakeholders do not yet perceive climate change as a problem for the management and conservation of livestock biodiversity.

In addition to the IPCC reports describing the predicted impact of climate change on ecosystems and agriculture (Easterling *et al.*, 2007), several papers provide a general overview of the expected impact of climate change on livestock production. Other papers model changes in production systems and species composition under climate change, poverty impact or projections of methane (CH_4) emissions from African livestock (Herrero *et al.*, 2008). De La Rocque *et al.* (2008) provide an overview on the impact of climate change on animal diseases.

The big mismatch between the low resolution of available data and the complexity of agricultural production systems makes it difficult to model the effects of climate change even for organisms with well-known environmental envelopes. Biogeographic models exist for the reaction of some well-described crop species to climate change. In the simplest case they project temperature and/or precipitation changes spatially and adjust the area under which a specific crop is able to produce. For some crop and forest species with known migration rates this allows the production of regional or national maps of projected species distribution. For livestock, such projections are more complicated: Firstly, several species were domesticated in the same region, implying that they have similar environmental envelopes. Secondly today, many breeds of the major livestock species are globally distributed, implying that the geographic distribution of specific breeds is overlaid by different production systems. In addition, detailed data on most breeds' adaptation traits, including their thermal neutral zones and spatial distribution, are not available (FAO, 2008b). Consequently, breed-level predictions or bio-geographic models of the implications of climate change on livestock diversity are hardly possible with current data.

Therefore, instead of trying to predict the survival or movement of specific breeds under climate change scenarios, this paper aims to shed light on the likely sensitivity of breed diversity to climate change and related drivers, the production and ecosystems breeds depend upon, and the goods and services they supply. For the purpose of this paper, a conceptual approach based on a simple overlap of breeds and production systems is used to identify possible implications of climate change for breed diversity.

Livestock Adaptation Differences Relevant for Climate Change Adaptation

Genetic mechanisms influence fitness and adaptation. Barker (2009) defined adaptedness as the state of being adapted, the ability of breeds to produce and reproduce in a given set of environments, or the choice of particular breeds for specific environments. Adaptability is then a measure of potential or actual capacity to adapt, for example if one breed is used in different environments. Adaptation traits are usually characterized by low heritability. In relatively stable environments, such traits have probably reached a selection limit; however, they are expected to respond to selection if the environment shifts, resulting in change of fitness profiles and increase in heterozygosity (Hill & Zhang, 2009).

Physiological Stress and Thermoregulatory Control

Heat stress is known to alter the physiology of livestock, reduce male and female reproduction and production, and increase mortality. Livestock's water requirements increase with temperature. Heat stress suppresses appetite and feed intake; thus feeding rations for high-performing animals need to be reformulated to account for the need to increase nutrient density. Body temperatures beyond 45– 47 °C are lethal in most species. Heat stress is an important factor in determining specific production environments already today (Zwald *et al.*, 2003). Temperature is predicted to increase globally, with reduced precipitation in many regions, particularly in already arid regions. While substantial differences in thermal tolerance lie between species, there are also differences between breeds of a species.

Ruminants generally have a higher degree of thermal tolerance than monogastric species, but species and breed environmental envelopes overlap. The ability to thermoregulate depends on complex interactions among anatomical and physiological factors. Factors such as properties of the skin and hair, sweating and respiration capacity, tissue insulation, the relationship between surface area per unit body weight or relative lung size, endocrinological profiles and metabolic heat production are known to influence heat loads, but the underlying physiological, behavioural or genetic mechanisms are largely unknown. With increasing milk yield in dairy cattle, growth rates and leanness in pigs or poultry, metabolic heat production has increased and the capacity to tolerate elevated temperatures has declined. In the long term, single-trait selection for yields will therefore result in animals with lower heat tolerance.

Measurement of the effects of heat and other stressors is difficult. The effect of heat stress on milk yield at specific test days is more immediate and easier to measure than on growth. After research on heat resistance of different species in the 1970s and 1980s, there is still today a lack of experimentation and simulation of livestock physiology and adaptation to climate change, which makes it difficult to predict impacts or develop adaptation strategies.

In addition to standard physiological measures of heat stress, such as rectal temperature or heart and respiratory rates, measurements of net radiation and convection are required to evaluate the implications of heat stress in extensive grazing systems. Heat tolerance tests will give misleading results unless modifying factors such as age, nutrition, state of health, reproduction and emotion, physical activity, level of production, acclimatization and management are taken into consideration (Bianca, 1961). However, research into behavioural or metabolic breed differences is in its infancy.

A wealth of literature is available on adaptation differences between zebu and taurine cattle. *Bos indicus* is generally more heat resistant than *Bos taurus* (Burns *et al.*, 1997), with zebu cattle maintaining lower rectal temperatures, lower respiration rates and lower water requirements than taurine breeds (King, 1983). However, research is biased towards few breeds. Many of these studies involved international transboundary cattle breeds, dairy more than beef. There are fewer breed-level studies of local breeds within the taurine or zebu cattle groups, and even fewer in other livestock species. In general, the high-output breeds originating from temperate regions that provide the bulk of market production today are not well adapted to heat stress. Milk production, fertility and longevity, in Holstein Friesian cattle for example, decline as temperature increases (West, 2003; St-Pierre *et al.*, 2003). Large White sows are less heat tolerant than Creole sows.

On the other hand, many species and local breeds, particularly those from the Near East and Africa, are already adapted to high temperatures and harsh conditions (FAO, 2006b). The distribution of some domesticated species is completely or mainly restricted to arid lands. Camelids are mostly found in arid areas, with the species differing in their adaptation to altitude and climatic zones. Yaks are adapted to very harsh high-altitude environments in the Asian drylands. More than 70% of breeds of ass, around 50% of sheep and goat breeds, and 30% of cattle and horse breeds reported are adapted to arid areas (FAO, 2006b). Most local breeds are, however, not well characterized and their adaptation includes not only heat tolerance but also to their ability to survive, grow and reproduce in the presence of poor seasonal nutrition as well as parasites and diseases. Breeds adapted to these dry areas will more likely be affected by natural resources degradation linked to climate change rather than temperature or precipitation change *per se*.

Nutritional Stress

Breed and species differences in diet selection in sheep and cattle have been observed, linked to different metabolic profiles. Hayes et al. (2009a) noted that there is more genetic variation in dairy cattle sensitivity to feeding level than to heat stress. Breeds differ with regard to their mobilization of body resources to cope with periodic underfeeding and cease reproduction at different levels of body weight loss. Also rumen physiology, the ability to walk and reach scarce feed resources, to intake water and rehydrate, or to respond with increased night-time grazing to high afternoon temperatures, or even genetic aspects of diet selection play a role (Hall, 2004). The relationships between energy reserves, endocrinological parameters and breeds' reproduction performance need further attention. Contrary to other species, feed intake and digestibility in dromedaries did not decline under heat stress. Among cattle, zebuine breeds generally deal better with lowquality forage than do taurine breeds, while taurine breeds have a better feed conversion ratio (FCR) with high quality feed. Livestock can compensate the expected climate change induced shrub encroachment to a certain extent if the animals are able to select high-quality diets from different plant components or species.

However, there is more anecdotal than scientific evidence for breed differences in feeding behaviour and browsing ability. For example, the West African Sokoto Gudali (Blench, 1999) and the South African Nguni (Bester *et al.*, 2002) cattle breeds seem well adapted to browsing. Merino and Dorper sheep, Welsh Mountain and Scottish Blackface sheep (Fraser *et al.*, 2009a, b), and Cashmere and Celtiberic goats (Jauregui *et al.*, 2008) differed in their grazing/browsing behaviour. More research on breed differences in feeding behaviour and adaptation to specific environments is therefore needed.

In addition to diseases affecting the animal itself, a new range of pests and diseases will impinge on crop and forage species, thus affecting the quantity and quality of livestock feeds.

Disease Stress

Climate affects vectors, pathogens, hosts and host-pathogen interactions from the level of cellular defence to that of the habitat. Hoberg *et al.* (2008) provide an overview of predicted responses of complex host-pathogen systems to climate change. Climate change may affect the spatial distribution of disease outbreaks, and their timing and intensity. Outbreaks of African horse sickness, peste des petits ruminants, Rift Valley fever, bluetongue virus, facial eczema and anthrax are triggered by specific weather conditions and changes in seasonal rainfall profiles. The predicted reduction in the availability and quality of water will increase the risk of water-borne diseases for humans and livestock.

Climatic effects on host-vector and host-parasite population dynamics will further the geographic expansion of vector-borne infectious diseases (e.g. Rift Valley fever, bluetongue and tick-borne diseases) to higher elevations and higher latitudes and affect the transmission and course of the diseases. However, expansion of the range of a pathogen or vector does not necessarily result in wider disease transmission (De la Rocque *et al.*, 2008). For some diseases, abundance of competent hosts and the effects of risk factors, such as the movement of animals and changes in production systems, habitats and ecosystems, will remain more important than climate change. The possible impact of bluetongue on endemic sheep breeds in the UK is raised by Carson *et al.* (2009).

Rapid spread of pathogens or even small spatial or seasonal changes in disease distribution may expose naïve livestock populations to new diseases. Such newly-exposed host populations lack resistance or acquired immunity; this may result in more serious clinical disease. The expected increased and often novel disease pressure will favour genotypes that are resistant or tolerant to the diseases in question. FAO (2007a) lists breeds, mainly from developing countries, that are reported to be resistant or tolerant to trypanosomiasis, tick burden, tick-borne diseases, internal parasites, dermatophilosis, or foot rot (59 cattle breeds, 33 sheep, 6 goat, 5 horse and 4 buffalo breeds). Again, many of these reports are based on anecdotal evidence rather than scientific studies, and the underlying physiological and genetic mechanisms are not well understood. Various studies have been undertaken to map genes and study gene expression in relation to these diseases, but no reports verifying causal mutations have been produced.

Climate Change Adaptation of Livestock Production Systems

The IPCC has defined "Adaptation" as "Initiatives and measures to reduce the vulnerability of natural and human systems against actual or expected *climate change* effects". There are many ways in which producers can adapt to climate change. Without judging possible differences in the efficiency of these measures, the paper focuses on the utilization of animal genetic diversity as one option for climate change adaptation. Producers can adapt to climate change by adapting their animals' genetics to the changed environment, or by adapting the production environment while maintaining the animal genetic portfolio. It is assumed that farmers will first use adaptation technologies that can be quickly deployed, and will only change their genetic resources portfolio when it becomes unavoidable. Climate change adaptation can be considered in two ways:

i)   How can animal genetic resources cope with and adapt to climate change while continuing to contribute to food security and rural livelihoods?

ii) How can the option value of genetic resources be maintained and potential loss of diversity minimized in the event of climate change?

Adaptation of Husbandry Practices

Frankham (2009) notes that the adaptation of high-output breeds to confined production environments is a recent phenomenon. The direct effects of climate change on housed livestock are expected to be small, as management can compensate for losses in animal fitness by modifying the environment. A variety of technologies can be used to deal with the effects of short-term heat waves, including shading or sprinkling to reduce excessive heat loads (Marcillac-Embertson *et al.*, 2009). Access to these technologies and to capital will determine the ability of producers to protect their herds from the physiological stress of climate change. Intensive livestock production systems have more potential for adaptation through technological changes and this may make them relatively insensitive to climate change and allow high-output breeds to be retained (Adams *et al.*, 1998). Freitas *et al.* (2006), for example, found that effects of heat stress were smaller in larger herds, which were more likely to afford high-efficiency cooling devices.

The widespread adoption of such technologies will also depend on the availability and prices of energy and water. The question is: How long can the production environment of high-output breeds be maintained in view of expected increases in feed, energy and water prices? In extensive or pastoral systems, where the rate of technology adoption is generally low, or in regions that already today have a limited capacity to adapt (e.g. sub-Saharan Africa), the risk of breed displacement or loss may increase. However, local breeds under traditional management are generally more resilient to environmental changes than are high-output breeds.

Species and Breed Shifts

Shifts Due to Agro-Ecosystem Changes

The use of multi-species and multi-breed herds is one strategy that many traditional livestock farmers use to maintain high diversity in on-farm niches and to buffer against climatic and economic adversities. Such traditional diversification practices are useful for adaptation to climate change. Seo & Mendelsohn (2007, 2008) modelled that small farms in developing countries were found more climate change resilient due to their more diverse species portfolios, the ease with which they can shift between species and diversify, and their reliance on goats and sheep. On the contrary, commercial dairy and beef operations were more vulnerable than small farms, because their specialized nature makes it difficult for them to switch to other species.

Several livestock species-level models that take into account the direct effects of climate change together with changes in agro-ecological conditions and production systems indicate that farmers will switch from cattle and chickens towards goats and sheep as temperature rises. The models agree that in Africa, livestock in semiarid rangelands will expand at the expense of humid and temperate/tropical highlands systems, they differ, however, in the relative share of these shifts.

Seo & Mendelsohn (2008) predict that ruminant numbers in rangelands will increase as long as there is sufficient precipitation to support vegetation growth. Jones and Thornton (2009) predict that livestock keeping will replace cropping in today's marginal mixed crop-livestock systems as they become ecologically and socially more marginal. In contrast, Herrero *et al.* (2008) model shifts of livestock populations from rangelandbased grazing to mixed systems based on improved feeding of crop by-products.

The outcomes of these models imply different breed portfolios, as local breeds are more adapted to pastoral systems. Species substitution due to climate and vegetation changes has already been observed in the Sahel, where dromedaries replaced cattle, and goats replaced sheep, following the droughts of the 1980s. In countries such as Niger and Mauritania, and in northern Nigeria, camel rearing is now a common activity. Unlike cattle and sheep, which largely feed on herbaceous vegetation, camels browse on shrubs and trees, while goats use both strata. The use of browsing species has several advantages: the browse strata cannot easily be used by other species and tends to offer green forage also during the dry season; and browse is increasing in some environments due to overuse of the herbaceous layer. While species and breed displacement from the arid and semi-arid to the sub-humid zones in West Africa has been observed, extension into the humid zones, where disease pressure is high, is still limited.

Climate Change Mitigation Measures

Despite contributing only 25% to GHG emissions from the livestock sector (FAO, 2006a), public discussion focuses on enteric fermentation in ruminants. Reducing livestock numbers, increasing individual animal resource use efficiency and optimization of feed rations and feed additives or other technologies may be primarily used to reduce CH_4 excretion in ruminants. In general, CH_4 output increases with the higher dry matter intake linked to high performance, however, the production pathways of different animal products differ in their GHG emissions and this may influence the future emphasis given to different production systems – and the related breeds. For beef cattle, intensive feedlot systems give rise to less CH4 per unit of meat produced than do extensive grazing systems, because CH4 decreases as the proportion of concentrate in the diet increases, and because of faster growth rate and shorter time to market.

Milk protein can be produced with less CH4 emission than beef (Williams *et al.*, 2006); CH4 emission per kg of milk declines as production increases, but with a diminishing rate. In a live-cycle-assessment of global GHG emissions per kg of fat and protein corrected milk, Gerber & Vellinga (2009) found that intensive and mixed farming dairy production had lower emissions than grassland based systems. Also, industrialized countries have lower emissions than developing regions. Improved genetic potential of the cattle, increased feed quality and manure management will reduce emissions from extensive systems. In an intermediate GHG reduction scenario, dairying might become the major focus of cattle production, and beef may become a by-product of dairying. Dual-purpose breeds and crossbreeding may gain importance.

Local ruminant breeds with their relatively lower output and higher GHG emission per kg of single food products are considered unproductive. However, the productivity equation should take account of the multiple products and services provided by livestock in most smallholder and pastoral production systems (Ayalew *et al.*, 2003). When considering GHG emissions from enteric fermentation, account should be taken not only of the gross efficiency of converting feed inputs to human food, but also of differences in species' ability to use forages that cannot otherwise be used by humans, historical and alternative herbivores (e.g. wild ungulates) and C-sequestration in grasslands which may partially offset GHG emissions from other components of the production process. Improved pasture management (e.g. restoring soil organic matter, reducing erosion, decreasing biomass losses resulting from burning and overgrazing) has positive environmental effects (soil C-sequestration, biodiversity) and a favourable impact on livestock productivity (Smith *et al.*, 2007). Gill and Smith (2008) propose using 'human edible return' as another indicator to assess livestock efficiency. This would favour the return of herbivore livestock species to foragebased feeding and land-based production systems and result in a different breed portfolio than the intensification pathway.

Breeding for Climate Change Adaptation and Mitigation

Although the direct effects of climate change on the animals are likely to be small as long as temperature increases do not exceed 3° C (Easterling *et al.*, 2007), projections suggest that further selection for breeds with effective thermoregulatory control will be needed. This calls for the inclusion of traits associated with thermal tolerance in breeding indices, and more consideration of genotype-by-environment interactions (GxE) to identify animals most adapted to specific conditions.

Breeding for climate change adaptation or mitigation will not be fundamentally different from existing breeding programmes; however, the problems related to measuring the phenotypes relevant to adaptation have to be overcome. In the past decade, breeding goals in many commercial breeding programmes have broadened through changes in selection indices and aim to improve production, longevity and functional traits simultaneously – for example in dairy cattle, pigs and layer chickens (Wall *et al.*, 2008). As a result, correlations between breeding values with broader indexes that include functional traits are lower than those with production traits only (Mulder *et al.*, 2006).

Correlations between the performance of genotypes in different environments are less than unity because of GxE, differences in trait definitions, and differences in data collection and analysis procedures (Zwald *et al.*, 2003). Such correlations between breeding values are lower in high temperature countries suggesting that heat stress plays an important role in GxE (Zwald *et al.* 2003). Biologically important GxE is assumed if the correlations between the performance of the same genotype in different environments are below 0.8 (Robertson, 1959). A single breeding programme with progeny testing of sires in different environments and applying index selection to simultaneously improve performance in those environments is recommended for genetic correlations between environments higher than 0.6. At lower genetic correlations between environments, environment-specific breeding programs are necessary to breed for special adaptability (Mulder *et al.*, 2006).

Heat Tolerance

Finocchiaro et al. (2005) proposed the use of heat-resistant individuals in a sheep breeding program as a main strategy to improve animal welfare and productivity in hot climates. Various physiological and blood parameters differ between local and exotic cattle breeds in Brazil (McManus *et al.*, 2008). Several Latin American cattle breeds with very short, sleek hair coat were observed to maintain lower rectal temperatures, and research in the major "slick hair" gene which is dominant in inheritance and located on Bovine Chromosome 20 is ongoing. Collier *et al.* (2008) suggest that there is some opportunity to improve heat tolerance through manipulation of genetic mechanisms at cellular level.

Selection for heat tolerance in high-output breeds based on rectal temperature measurements and inclusion of a temperature-humidity index (THI) in genetic evaluation models are promising. Different parameters, such as THI or dry-bulb temperature measurements, are used as indicators for heat stress. Different THI definitions were found prefable in the US, depending on the extent of natural and artificial evaporative cooling. The genetic variance due to heat stress was substantial at high THI.

However, in the dairy sector it may be difficult to combine the traits desirable for adaptation to high temperature environments with high production potential, because there seem to be different physiological and metabolic processes controlling heat tolerance and milk yield on the one hand, and heat tolerance and reproductive performance on the other. In beef cattle, the genetic antagonisms between adaptation to high temperature environments with high production potential seem to be more limited than in dairy, and improved characterization of adaptive traits, use of reproductive technologies and molecular markers, and strategic crossbreeding are being incorporated into programmes, for example in the Australian beef Cooperative Research Centre (Prayaga *et al.*, 2006).

Productivity and Feed Efficiency

Increasing productivity is a condition *sine qua non* for all production systems because of the need to make efficient use of the available inputs and to reduce the livestock sector's environmental footprint. In addition to selection for increased production *per se*, any selection that reduces mortality and increases early maturity, fertility and longevity tends to contribute to reducing GHG emissions per unit of output. Breeding for high performance and improved FCR, and reduced mortality due to better hygienic management, have significantly reduced the amount of feed (and land needed to produce this feed) per unit of product – more in monogastrics and in dairy cattle than in beef cattle or sheep. Jones *et al.* (2008) investigated the role of genetic improvement in a Life-Cycle-Analysis model in the UK and found that the annual reduction in GHG emission ranged from 0.8% in pigs and dairy cattle to 1.2 and 1.3% in broilers and layers, respectively. The largest contributions in broilers came from improved FCR, and in pigs from improvements to growth rate and fertility. Genetic gain in milk performance has considerably reduced the environmental impact of diary production in the USA (Capper et al., 2009). Future options for selection in ruminants lie in the host components of rumen function, in post-absorption nutrient utilization and in disease resistance. In pigs and poultry, the genetic variation in digestion parameters can be exploited (Warkup, 2007).

In addition to potential for fertility improvements in ruminants – for example decreasing the age of first-calving in zebuine cattle – there is sufficient genetic variability in feed intake, independent of liveweight and daily gain (Flachowsky & Brade, 2007) to permit selection for this trait. Assuming that future dairy systems may become more reliant on pasture than grain feeding, Hayes *et al.* (2009a) proposed to select sires whose daughters will cope better with low feeding levels and higher heat stress. They identified markers associated with sensitivity of milk production to feeding level and sensitivity of milk production to THI in Jersey and Holstein. Because feed-efficient animals are also more cost-effective and productive, the Australian beef industry now includes net feed efficiency as an integral part of its breeding programme (Beef CRC). Alford *et al.* (2006) calculated that CH4 could be reduced by up to 16% in 25 years if residual feed intake (RFI) were included in beef selection programmes. Initial costs to identify individuals with low RFI are high, however, particularly in grazing animals (Arthur *et al.*, 2004). Because of the above-mentioned differences in feed quality, productivity improvements in pasture-fed ruminants in the tropics will result in higher relative CH4 reductions than in ruminants grazing more digestible temperate pastures (McCrabb & Hunter, 1999). Possible synergies between plant and animal breeding need to

Disease Resistance

Experiments in domestic species have shown that there are often genetic differences in responses to disease challenge (Bishop *et al.*, 2002). Some of this variation is caused by single genes and some by multiple genes each with small effect. There is potential for genetic improvement of disease resistance, and various commercial breeding programmes already include resistance against helminthosis, ticks, mastitis, *E. coli* or scrapie. Extensive research on the genetics and breeding for worm resistance has been carried out in Australia, New Zealand and recently also in South Africa (de Greef, 2009). Breeding for disease resistance depends very much on the type of disease and the hosts' resistance or tolerance mechanisms, the availability and costs of alternative treatment (e.g. vaccines, drugs) and antimicrobial resistance of pathogens. In any case, the importance of molecular markers and marker assisted selection in such breeding programmes will increase.

Challenges

The speed of artificial selection depends on many genetic factors, among others on reproductive technology, selection procedures and on the accuracy of phenotyping. Breeding for improved performance has become a high-tech exercise; the technologies and skills required present a bias towards certain breeds and production systems. Similarly, while GxE is the measure of choice for assessing variability of breed performance and adaptation across different environments, there are several caveats related to its wide application.

1. Limited breeds: On a routine basis, publicly accessible GxE across countries is only estimated for sires within 6 international dairy cattle breeds, through the international genetic evaluations performed by Interbull that treat sire's multiple trait performance in each country as a different trait. Routine GxE assessments are not available for other cattle breeds or species.

2. Limited countries: Interbull's customers are mainly based in developed countries; South Africa is the only developing country customer. GxE effects are more pronounced if tropical countries are considered (Ojango & Pollott, 2002).

3. Limited production systems: Current AI bull evaluations are mainly based on daughters producing in high-input production systems and often do not differentiate between environments within countries (Zwald *et al.*, 2003). Even within the US dairy industry, GxE exists between husbandry systems and climatic zones, e.g. between grazing and confined dairy herds, especially those in extensive, hot areas (Kearney *et al.*, 2004).

4. Demanding data quality and analysis: Only a small percentage of national herds are usually used for progeny testing. Electronic

data capture which increasingly forms the basis of genetic evaluations is mostly prevalent in large herds; in future even fewer herds will be needed for progeny testing. Adaptation traits are more difficult to study and to record, have lower heritability, higher levels of non-additive genetic variation and phenotypic variance, and are more susceptible to GxE than production traits (Frankham, 2009). Even in dairy cattle populations in some developed countries, female reproduction traits are incompletely recorded (Goddard, 2009) – not to mention other functional or adaptation traits.

5. Economic constraints: Although the genetic correlations between developed and developing countries are probably lower than 0.75 (Ojango & Pollott, 2002) – a threshold above which it is genetically reasonable to import semen from large breeding programs (Mulder *et al.*, 2006) – there are no breeding programmes in developed countries that target developing country environments. As only a relatively small amount of genetic material is sold to developing countries, commercial breeders find it hard to justify specific breeding programs for such environments.

The majority of developing countries import genetic progress in production traits rather than developing it in their local breeds, as was highlighted in the low number of breeding programmes reported in the FAO State of the World report (FAO, 2007a). Gollin et al. (2008) found that the more economically advanced developing countries are importers of genetic material, while the poorest countries are not engaged in any international trade in animal genetic resources. Only a few countries with well-developed breeding institutions, research, extension and artificial insemination services have commercially relevant tropical cattle breeds, tropically adapted taurine, zebuine or composites and even fewer countries have commercially significant breeding programmes for adapted breeds of the other species (Madalena, 2008). This may strengthen the market position of commercial transboundary breeds in developing countries and indirectly continue to preclude characterization and selection within local breeds from these countries for increased production or even improved adaptation.

Genomic selection may exacerbate these trends, through the related requirements of reference populations for phenotyping. The accuracy of genomic breeding value estimation depends on the number of animals with phenotypes and genotypes in the reference population from which the SNP effects are estimated. Genomic selection is being applied in dairy breeding where consortia of main breeding companies are forming to assure the high numbers of progeny-tested bulls needed in the reference population. However, it is more difficult to implement in beef cattle or other ruminant species due to the dispersed breeding structure. In poultry, genomic selection is being tested across populations within companies (Avendano, 2009). A minimum of 500 progeny tested sires, with genotypes and full phenotypes recorded, are needed; this number can be reduced to 200-300 in breeds where the current genetic trend is close to zero (Ducrocq, 2010). Although costs for highthroughput genotyping have dropped considerably, genotyping and the phenotypic characterization and bioinformatics tools needed for their calibration are still most likely to be used in developed countries.

Genomic selection is promising but not yet transferable to developing countries where structured phenotyping and performance recording are largely missing and reference populations would be difficult to establish even if the genotyping could be sourced out. In order to be transferable across breeds or crossbreds, very high density SNP panels need to be developed and reference populations should include at least some individuals from all target breeds. The forthcoming 600K bead chip and improved imputation techniques will allow for better multi-breed analyses (Ducrocq, 2010). Also, GxE may reduce the accuracy of predicted genomic breeding value (Hayes *et al.*, 2009b) because linkage disequilibrium increases in adaptation to new environments (Frankham, 2009). Lillehammer *et al.* (2009) have developed a framework for evaluating effects of SNP in the presence of GxE, but this approach needs testing across a wide range of environments.

Conservation and Exchange

Conservation measures for threatened breeds have already been established in some countries (FAO, 2007a) and are a priority of the *Global Plan of Action for Animal Genetic Resources* (FAO, 2007b). The IPCC predicts an increase in disturbance and catastrophic weather events. Loss of animals as a result of droughts and floods, or disease epidemics related to climate change may thus increase (FAO, 2008a). If breeds are geographically isolated (endemic) – as is the case for some local and rare breeds – there is a risk of their being lost in localized disasters (Carson *et al.*, 2009). To secure against such disasters, it is necessary to characterize animal genetic resources and subsequently to build inventories, including information on the spatial distribution of breeds and valuable breeding stocks.

This may include precautionary cryo-conservation of genetic material, or other measures to ensure genetic recovery following a disaster. In the field of nature conservation and for crop wild relatives it is now argued that *in situ* strategies have to account for the fact that conditions in species' historic ranges will change, and indeed are already changing (McCarty, 2001; Jarvis *et al.*, 2008). Further to claims to facilitate species migration and maximizing adaptation opportunities through the maintenance of intact ecosystems, this results in a review of *in situ* conservation. However, recognizing that climate change may affect our food system quickly, the authors stress the urgent need to identify priority core species for collection and their inclusion in genebanks.

Similarly in animal genetic resources conservation, the focus may shift from *in vivo* to *in vitro* conservation. Most conservation programmes are based in developed countries with strong collaboration between genebanks and the animal breeding industry (FAO, 2007a). In developing countries, few breeds of the five major species are covered by conservation programmes, and programmes are of variable quality; the focus is typically on *in vivo* conservation (FAO, 2007a). Consequently, the sensitivity of *in situ* conservation programmes to the effects of climate change should be assessed and *ex situ* conservation measures be taken if needed. The role of the private genebanks held by breeding organizations or companies, and that of public genebanks still needs to be defined, but could involve the establishment of genebanks for local breeds and backups for commercial breeds. The operational protocols (e.g. material transfer agreements) remain to be developed.

Exchange

Most livestock production systems depend on species originally domesticated elsewhere and breeds developed in other countries and regions, making most countries highly interdependent with respect to animal genetic resources (FAO, 2007a,b). Climate change will increase the need to maintain wide access to animal genetic resources in the interests of future food security. Livestock breeding and production systems are complex and knowledge intensive. New species or breeds may replace the current ones as single new components in a production system or they may be changed together with other components of a system, including knowledge components. In human-managed systems, "establishment" of new species or breeds depends on how many components of the old production system can be transferred to the new area/system and on the socio-economic conditions. Blackburn & Gollin (2008) emphasize that successful introduction of new breeds into the USA has been based on several production traits and the interest and acceptance of the private sector, while introduction to take advantage of single traits has not proved sustainable, especially when other economically important traits were compromised. In any case, such replacement process may involve considerable costs and substantial investments in learning and gaining experience.

Although specific traits of tropical breeds may become important, it remains to be seen whether the impact of climate change will lead to a reassessment of the value of local breeds. The arguments Hill & Zhang (2009) make to explain why little use is made of conserved lines or breeds of developed countries to increase adaptation in commercial populations equally apply for local populations of developing countries: the production performance differentials between commercial, intensively selected breeds and any local breeds are so huge that selection for improved adaptation within those breeds offers far more opportunity than crossing or introgression of adapted genes.

This is even more important as genomic selection will make better use of the genetic variation present in commercial populations. In any case it can be expected that only well-characterized breeds will be used for targeted crossing or gene insertion to increase the adaptedness of highoutput breeds. However, most developing countries have insufficient resources or capacity for genetic or phenotypic breed characterization (FAO, 2007a). Only if climate change exceeds the adaptive capacity of the currently used genetic portfolio, would countries need to depend on better-adapted genetic resources from other countries to adapt their food and agriculture systems.

In this case, increased strategic crossbreeding with better adapted breeds, or insertion of specific genes through the use of biotechnology, may occur (McManus *et al.*, 2008). The importance and value of specific genetics would thus increase. Such changes in the species or breed mix may potentially lead to a reverse in the current flow of genetics. Countries that happen to host sought-after resources may then try to take advantage of their scarcity and control access to what will have become crucial genetic resources. The need for improved exchange mechanisms for animal genetic resources and the associated knowledge would thus increase as well.

Climate change is one additional factor affecting the already highly dynamic livestock sector. However, due to its slow but long-term effect and more pressing current needs such as increasing demand for animal products, climate change is not yet fully on the radar screen of the livestock community. It will increase the need for resource-efficient livestock production and may thus intensify current trends with a growing dichotomy between livestock kept for livelihoods by smallholders and pastoralists, and those kept for commercial production. The direct effects of climate change depend very much on the production and housing system, resulting in a buffered effect for the high-output breeds in confined systems.

Climate change mitigation measures in conjunction with the "traditional" economic drivers may have implications of the breed portfolio. Measures that reduce the land area previously used for grazing may add to the threats to local ruminant breeds and affect the livelihoods of their keepers. Changes in the land area devoted to cropping (including fodder) relative to rangeland and their relative productivity will influence the balance between non-ruminant and ruminant production, as will GHG reduction targets. In general, superior FCR will grant monogastrics a comparative advantage over cereal-fed ruminants. Efficiency differences between breeds will influence the proportion of commercial versus local breeds. High-output breeds of all species selected for improved FCR and high yield will dominate the production of milk, eggs and meat. These breeds will continue to out-compete local breeds. FAO (2007a) indicates that breed loss and risk in the past century was highest in regions that have the most highly-specialized livestock industries and in the species kept in such systems. The threat of extinction for local breeds, especially of monogastric species will increase. However, the outcomes of different scenarios and models imply different species or breed portfolios, making it difficult for decision makers to make rational choices.

Although many existing technologies in animal genetic resource characterization, conservation and breeding will be crucial for climate change adaptation and mitigation, research gaps exist, especially with regard to the physiology and genetics of adaptation. Research consortia such as GLOBALDIV have an important role to play in this regard. Currently, most research on the genetic variability of adaptation traits is undertaken with high-output breeds. There is need for long-term comprehensive breed characterization studies to shed light on the biological basis of adaptive traits. At current state of knowledge, it is not predictable whether climate change will be faster than natural or artificial selection. Developed and developing countries differ with regard to their portfolios of genetic resources and the management of these resources. Most tropically adapted breeds reside in developing countries and are largely uncharacterized and without structured breeding or conservation programmes. On the other hand, most high-output breeds are selected for the requirements of developed country markets and production systems.

Without support in breed characterization, breeding and conservation for developing countries, the divide between the scientific "haves" and "have-nots" will increase (OECD-FAO, 2009). Also the performance differentials between local breeds and high-output breeds, the long-term commitment required for genetic improvement and the ease of genetic material imports may discourage developing countries from initiating their own breeding programmes.

However, for the optimal utilization of the adaptation traits harboured in all breeds, research in genetic characterization and understanding adaptation in stressful environments needs to be strengthened. In view of the uncertainty for future developments, the use and non-use values of animal genetic resources should be maintained. Agricultural biodiversity must be an integrated part of climate change adaptation and mitigation efforts. Especially in marginal areas, climate change adaptation, biodiversity conservation and poverty alleviation should complement each other. Strategies thus need to be developed that strengthen livestock keepers' adaptation strategies, their ecological knowledge and local institutions.

The recent adoption by the international community of the *Global Plan of Action for Animal Genetic Resources* and the *Interlaken Declaration* provides for the first time an internationally agreed framework for the management of animal genetic resources, and also propose measures to support developing countries in their endeavours (FAO, 2007b).

2

Genomic Studies of Domestic Animals

The first decade of the 21st century has been a golden time for the advancement of genomics, driven by the completion of the Human Genome Project (HGP). Various methodologies and technologies have been developed during and after the process of building the human genetic blueprint that have been directly transferred into the studies of domestic animal genomics. The search for genetic underpinnings of human diseases perplexed researchers for many years. Only recently did the genetic factors underlying various human diseases begin to be revealed, especially with the help of genome-wide association studies (GWAS) using SNP arrays.

Single nucleotide polymorphisms (SNPs) are bi-allelic genetic markers, and they are easy to evaluate and interpret and are widely distributed within genomes. With proper coverage and density over the whole-genome, SNPs could capture the linkage disequilibrium (LD) information embedded in the genome, which could be used to pinpoint genes underlying human diseases. For domestic animals, these tools can contribute to i) better understanding of species' evolution, domestication and breed formation, and developing new theories of population genetics; ii) dissecting the genetic mechanisms of complex agricultural traits; and iii) improving selection methods for genetic improvement of animal production.

High-density SNP arrays were built for important farm animals, first for those with reference genomes and then recently also for others without reference genomes with the advent and application of the massive parallel sequencing technologies. The preparation and utilization of SNP arrays are having considerable impacts on the theory and practice of animal breeding and genetics, which will play important roles in the years to come.

In this review, the whole-genome sequencing and HapMap studies of several important domestic animals are briefly summarized. Then, details about the development of SNP arrays and their applications in various genetics and genomics research projects are also reviewed. Lastly, lessons learned from the reported studies and prospects for future work are discussed.

Animal Genome Sequencing and Hapmap Projects

The whole-genome sequencing strategies for most domestic animals were taken directly from human genome sequencing, i.e., combining both whole-genome shotgun (WGS) and BAC-to-BAC sequencing (Green, 2001). Based on their significance in agriculture and as biomedical models, chickens, dogs, cattle, horses and pigs have had their genomes sequenced, as well as some other important animals. Due to the rapid development of "next-generation" sequencing technologies and the availability of reference genomes, these strategies have been modified for different species. For those well-studied species in which high levels of genomics knowledge and sequence coverage are required, such as chicken, dog, cattle, horse and pig, the dual approaches of WGS and BAC sequencing were applied.

These sequences provide comprehensive information for comparative genomics studies on the evolution and function of important genes and genomic regions. The comparative studies among the genomes of human and domestic animals have also demonstrated a high level of conservation and orthology for protein coding genes. However, huge differences were found in non-coding regions, especially intergenic repetitive regions, which may be one of the major forces driving evolution. The HapMap studies also revealed abundant genetic variability within and between domestic breeds. The majority of the variation was discovered by large-scale genotyping of SNPs and insertions or deletions of DNA fragments with variable sizes, such as copy number variation (CNV), which could in part contribute to the phenotypic diversity of domestic animals.

The additional information derived from linkage mapping, radiation hybridization (RH) mapping, fluorescent in situ hybridization (FISH) mapping and expressed sequence tags (ESTs) were used to assist in the genome assembly and annotation. For those with 'light' coverage genomes, such as dog (1.5x) and cat (2x), WGS with "next-generation" sequencing technologies were utilized, and the sequences were assembled using human and other closely-related species as references. More recently, due to the reduced cost of sequencing, both deep-sequencing and individual genome sequencing have been attempted in the cow and chicken, along with the 1,000 Genomes Project in human.

With the completion of whole-genome sequencing of domestic animals, HapMap projects were developed. Since domestic animals have rich sources of phenotypic diversity, which can be interrogated by SNPs across the genome, HapMap studies can be helpful to characterize the complexity of a genome and disentangle the genetic bases of complex traits.

The advantages of the HapMap studies include i) production of a large number of SNPs for design and preparation of high-density SNP arrays; ii) clarification of the genetic relationships among diverse breeds and the phylogenetic relationships between domestic animals and their wild ancestors; iii) prediction of the potentially significant historical events that occurred during domestication and breed formation, such as bottleneck effects and selective sweeps; and iv) identification of the potentially important candidate genomic regions associated with distinct morphology, disease and other quantitative traits. The implications of HapMap studies will be demonstrated in the later sections of this review.

Design and Development of High-density SNP Arrays

SNP Discovery

A very large number of SNPs are essential for the design and construction of arrays. Different methods and resources can be used for SNP discovery, including analysis of predicted SNPs generated from genome sequencing and HapMap studies, completing reduced representation library (RRL) sequencing, downloading SNP information from SNP of NCBI or collections of SNPs from individual research institutes or lab groups.

The completion of whole-genome sequencing and HapMap projects uncovered a large number of genetic variants across the genomes of domestic animals, most of which were SNPs. In the chicken, ~2.8 million SNPs were identified (International Chicken Polymorphism Map Consortium, 2004). There were more than 2.5 million potential SNPs in the dog genome, with one SNP per 0.9 kb between breeds and one SNP per 1.5 kb within breeds (Lindblad-Toh et al., 2005). In cattle, ~2.2 million draft SNPs were detected with one SNP per kb (The Bovine HapMap Consortium, 2009). In the horse genome, ~1.1 million draft SNPs were discovered with one SNP per 2 kb (Wade et al., 2009).

Although many SNPs were predicted during genome sequencing projects, SNP prediction software could confuse sequencing errors with true SNPs, meaning further validation is needed. The candidate SNPs for array design should be validated and have high minor allele frequency (MAF) in the testing populations. Matukumalli et al. (2009) found an uneven distribution of SNPs across the genome based on an analysis of cattle draft SNPs. Additionally, it was determined that the nucleotide conversion rate of SNPs was usually low and MAF was not estimated accurately because of the limited sample size of animals used in the HapMap studies. In the commercially released BovineSNP50 array, around three-fifths of SNPs were from the filtered draft SNPs from genome sequencing. Kerstens et al. (2009) used a similar pipeline to obtain 104,525 SNPs from 1.2 Gb of draft swine genome sequence and verified the polymorphisms of 134 from 163 filtered SNPs in several tested pig populations.

Another effective approach to identify large numbers of candidate SNPs is RRL sequencing, which was first introduced for creating a human SNP map (Altshuler et al., 2000). This approach could reduce the complexity of the genome by several orders of magnitude, help discover SNPs that are extensively dispersed across the genome, and can even be performed without a priori knowledge of the genome sequence.

Due to each species' unique genome sequence, the most suitable restriction enzymes for RRL sequencing are variable. In the human genome, BglII cut sites are commonly distributed (Altshuler et al., 2000). Van Tassell et al. (2008) constructed several RRLs that were generated from DNA of eight commercial dairy and beef breeds that were digested by the HaeIII restriction enzyme, and identified 62,042 putative SNPs by deep-sequencing and filtering procedures. Around two-fifths of the SNPs on the BovineSNP50 array were from the RRL sequencing approach (Matukumalli et al., 2009). In pigs, Wiedmann et al. (2008) identified 115,572 putative SNPs by sequencing of RRLs that were built from seven predominant commercial pig breeds and were digested by HaeIII. Amaral et al. (2009) also detected 17,489 pig SNPs using RRLs sequencing of pools of DNA from five Large White x Pietrain crossbred boars digested by the DraI enzyme. Ramos et al. (2009) prepared 19 RRLs derived from four popular commercial pig breeds and a wild boar and three restriction enzymes (AluI, HaeIII and MspI), and eventually obtained 372,886 high-confidence SNPs. In total, the SNPs obtained from the RRLs sequencing comprised about 94% of the 64,232 SNPs used in the commercially released PorcineSNP60 array (Ramos et al., 2009).

Illumina's Select Technology

Illumina's BeadArray based on single-base extension or allele-specific primer extension and Affymetrix's GeneChip based on molecular inversion probe hybridization are the two biggest and most competitive SNP chip genotyping platforms. The approaches of both arrays are different, but they have the capability to perform high-throughput genotyping for large scale samples. In comparison to the GeneChip, the BeadArray is cheaper and more flexible on probe designs (Perkel, 2008). Currently, the majority of the commercially released SNP arrays for domestic animals are constructed using the BeadArray platform with Illumina's iSelect Infinium technology.

A bead chip is a micro-electro-mechanical system, in which wells attaching the beads are created by combining photolithography and plasma etching on silicon wafers. The beads are randomly dispersed and assembled into wells on a silicon wafer (Steemers and Gunderson, 2007). The location of each bead on the array can be identified through a decoding process that uses a 29 base tag sequence linked to the bead. Each bead has a number of 50-mer locus-specific primers following the tag sequence, which are used to anneal the genomic sequences flanking the target SNPs.

After direct hybridization of the genomic DNA to the SNP array probes, each SNP locus is scanned by an enzymatic-based extension assay using fluorescent labelled nucleotides. The labels are visualized by staining with an immunohistochemistry assay to increase the signal intensity (Steemers and Gunderson, 2007).

The two different primer extension assays are allele-specific primer extension (ASPE) and single-base extension (SBE), which are called Infinium I and II assays, respectively. Infinium II can reduce the required number of synthesized beads by nearly half compared with Infinium I, and thus make this bead chip more economical. Therefore, the probes for the majority of SNPs in the chip follow the Infinium II design.

Criteria for SNP Selection

Whole-genome sequencing and HapMap projects provided each species with draft SNPs. High quality control (QC) criteria were then set up to filter these draft SNPs, which required i) each allele of the SNP is included in at least two sequence reads; ii) no repetitive elements surrounding the SNP (within 100 bp); iii) the SNP must be predicted by a minimum of six sequence reads; and iv) predicted SNPs cannot overlap with complex regions (e.g. duplicated sequences) (Matukumalli et al., 2009).

Candidate SNPs following this preliminary step were considered for placement on the SNP array. A set of additional criteria were set up, such as SNP distribution across the genome and each SNP's properties. The physical distribution of SNPs evenly across the genome and reasonable intervals between neighbouring SNPs (except Y or W chromosomes in mammals and birds, respectively) were prioritized. As far as the properties of each SNP were concerned, high MAF determined from sequencing of representative samples, high quality score and good validation status were required as well as the design score of Illumina's assay design.

In addition, the bead density, the redundancy of beads per bead type and the final expense (cost-effectiveness) influence the number of SNPs being assembled into the commercially released SNP arrays (Steemers and Gunderson, 2007). For several important farm animals including horse, cattle and pig, the number of SNPs in the first generation chips was slightly more than 50K, ensuring that at least 50K SNPs would work. The first version of the canine SNP array contained 20K, and now a high-density chip with 170K has been released. A high-density 500K cattle SNP array is also being developed (http://www.illumina.com).

Applications of the Snp Array

Genomic Selection

Genomic selection (also termed as genomic prediction or genomic evaluation) perhaps is the most fundamental change to breeding and genetics in agriculture as a direct result of the application of SNP arrays. It is different from human studies which have mainly concentrated on searching for disease genes or genealogy. Genomic selection is an advanced form of marker assisted selection (MAS) which concentrates on all markers across the whole genome (Meuwissen et al., 2001; Goddard and Hayes, 2009; Calus, 2010). The MAS strategy has been advocated for the past two decades, but its utilization is limited in practice because of lack of genetic markers linking to QTL with significant effects and the expense of genotyping. Meuwissen et al. (2001) proposed the original concept of genomic selection, i.e., predicting breeding values of animals using information offered by thousands of SNPs across the genome (genomic estimated breeding value, GEBV), by assuming the availability of abundant SNPs scattered throughout the genome and LD relationships between SNPs and QTL. With the new SNP arrays, more SNP effects need to be predicted than there are phenotyped animals for use in predicting these effects. Consequently, Bayesian analysis methods were initially tested to address this problem. Two kinds of Bayesian approaches were developed to predict GEBV using dense SNPs in this pilot study. In both Bayes models, the effect of each SNP was considered to be independent and random, and the variance of SNP effects were either assumed to be constant or locus specific, and then SNP effects were estimated by a Bayesian procedure with a prior distribution for this variance. These Bayesian methods had higher prediction accuracy compared to those of least squares (LS) and conventional best linear unbiased prediction (BLUP) based on simulation data. In recent years, different statistical approaches for genomic selection have been developed, derived from either nonparametric Bayesian models or parametric methods including genomic best linear unbiased prediction (GBLUP) and mixed regression models.

Based on most of the studies using simulated data, several major factors which influence the accuracy of genomic selection were recognized: i) the LD extent between SNPs and the QTL; ii) the size of the training population (the individuals both phenotyped and genotyped for building statistical models and predicting SNP effects); iii) the heritability or genetic basis of the analysed trait and iv) the distribution of QTL effects. Meuwissen et al. (2001) found that the prediction accuracy of genomic selection could reach 85% when [r.sup.2] between the adjacent SNPs was greater than 0.2, which required that SNPs have high-density and even distribution across the genome. For those traits with low heritability and composed of many QTL with small effects, a larger training population is necessary (Goddard and Hayes, 2009; Calus, 2010). Given the heritability of the targeted traits and the prediction accuracy in genomic selection programs, the numbers of animals required in the training population can be estimated.

Additionally, other methods have been proposed to improve the accuracy of genomic selection, such as estimation of missing genotypes, distinguishing the actual SNPs in LD with QTL from those only tracing relationships between animals, and developing novel approaches considering dominance and epistatic effects. Due to consistency of LD across populations, the prediction accuracy will likely remain at a high level whenever the training populations are at least partially related to the validation populations (animals for which GEBVs are being predicted without phenotype data), suggesting that SNP effects obtained from crossbred populations are suitable for genomic selection in pure breeds.

Another challenge is to carry out genomic selection in livestock across both national and global regions. Genomic selection has been adopted for genetic evaluation of dairy cattle in the United States of America (VanRaden et al., 2009) and is being considered in the International Cattle Genetic Evaluation Project.

With the availability of higher density SNP chips which can help find more common haplotypes between breeds, the improvement of advanced statistical approaches and computer programs, and joint sharing of phenotypes and SNP genotypes among research groups, breeding companies will be able to apply genomic selection for livestock across the globe (VanRaden and Sullivan, 2010).

Genome-wide Association Studies

Both candidate gene and QTL mapping strategies have been extensively utilized in domestic animals for the discovery of genetic markers suitable for MAS. However, the limitations of these approaches are becoming apparent. The biological mechanisms of quantitative traits and diseases are complicated, and they are still being explored. The determination of candidate genes according to their putative physiological roles is often difficult, and the candidate gene approach may miss the identification of novel genes and pathways associated with some traits. The regions with identified QTL are generally large and further fine mapping is necessary, and often consistency of results from QTL mapping is limited among different resource families (Rothschild et al., 2007). GWAS (also termed as whole-genome association studies, WGAS) is one of the most promising approaches to overcome these limitations.

Although GWAS have been carried out in domestic animals using the commercially available SNP arrays, most of them were on disease related traits because case-control study strategies could be easily utilized for association analyses. For quantitative traits such as growth rate, lean meat percentage, intramuscular fat content and milk production, some researchers tried single marker mixed model or mixed regression models for association analyses. Other researchers have used the posterior probability that is derived from a Bayesian approach originally designed for genomic selection, where the SNPs having highest posterior probability (i.e., the frequency of a SNP included in the model for GEBV prediction) are most likely to be linked to the QTL. A number of GWAS in several important domestic animals have been completed to date with significant results.

Whole-Genome LD Patterns

Construction of high-resolution LD maps, calculation of the extent of LD at the population level, and characterization of haplotype block structures are crucial for fine mapping and genomic selection. In most cases, the extent of LD between loci varies between populations, including lines, breeds and even different populations within a breed, and this inconsistency between groups of animals may have a significant impact on fine mapping, genomic selection and GWAS.

The extent of LD in a population also plays an important role in helping a researcher to decide the SNP density needed for a particular study. Differences in population structure and evolutionary forces affect how much LD exists in a population. For populations with longer range LD, there is less value in moving to a higher density SNP array because most QTL may already be in LD with markers on a smaller array. If LD has a relatively short range, then not all QTL may be in LD with markers on a smaller array such that use of a larger SNP array may be worth the extra cost. The extent of LD in a given population can be easily calculated from any SNP array study to predict the best array size to use in future studies.

The findings from the cattle and horse HapMap projects have demonstrated that the decay of LD relationship between SNPs slows beyond 100 kb, and haplotype blocks become smaller between breeds. It has been suggested that ~100K SNPs may be sufficient for association mapping within and across breeds.

Additionally, effective population size could be derived from the extent of LD within a given interval length, $[r.sup.2] = 1/(4Nc+1)$ or $[r.sup.2] = 1/(4Nc+2)$ (when mutation is considered in the model). Where N is the effective population size $1/2c$ generations in the past, and c is the recombination rate based on the number of Morgans between the examined markers (Sved, 1971). Although values of N varied between populations, rapid decreases in N were observed in recent generations in the populations examined, which implied that domestic animals have undergone inbreeding and extensive selection in the past two centuries, both well known occurrences.

In general, the amount of LD between any two markers decreases as the physical distance between those markers increases. Forces such as selection, however, can cause markers that are far apart physically (or even on different chromosomes) to be in high LD with one another. Having high LD for long stretches or between unlinked markers complicates fine mapping. One feasible approach for discovering SNPs that are widely applicable for selection is to carry out LD mapping in multiple breeds, so the SNPs in high LD with QTL across populations can confirm the associations.

Population Genetics

Selective sweeps : During domestication and breed formation of domestic animals, they have experienced both natural and artificial selection. These selection pressures have led to increased allele frequencies of some mutations in a few specific genomic regions because these mutations made the animals more adaptable or gave them favourable characteristics based on human demands. Over time other polymorphisms may have decreased in frequency or vanished, and a single haplotype containing multiple genes may have become the only one or the most prominent in the population. This has been termed as a selective sweep or positive selection (Andersson and Georges, 2004).

Several statistical methods were proposed for detecting selective sweeps (Sabeti et al., 2007). The integrated haplotype score (iHS) developed from integrated extended haplotype homozygosity (EHH) detects selective sweeps by identifying genomic regions with increased local LD. The fixation coefficient (Fst) can be used to predict selective sweeps by comparing the Fst values among populations. The composite likelihood ratio test (CLR) is based on the comparison of the maximum composite likelihoods under models with and without selective sweeps.

The above methods have been utilized for selective sweep detection in the Bovine HapMap Project (The Bovine HapMap Consortium, 2009). Based on the iHS method, specific haplotype frequencies in the genomic regions containing MSTN (relevant to muscle development),

ABCG2 (relevant to milk yield and composition) and KHDRBS3 (relevant to intra-muscular fatness) might have resulted from selective sweeps. Genomic regions relevant to behaviour, immune response and feed efficiency were discovered based on [F.sub.st] estimates (The Bovine HapMap Consortium, 2009). Both iHS and CLR approaches have revealed that one region including SPOK1 was subject to a selective sweep in beef and dairy cattle. A total of 12 putative selective sweep regions associated with residual feed efficiency, beef yield and intra-muscular fatness were discovered when additional data sets were included (Barendse et al., 2009). In addition, a set of genes including GHR, MC1R, FABP3, CLPN3, SPERT, HTR2A5, ABCE1, BMP4 and PTGER2 were possibly subject to selective sweeps.

In the previously described chicken HapMap Project, most SNPs were thought to arise before domestication. However, Rubin et al. (2010) using massively parallel sequencing identified a possible selective sweep resulting from domestication and specialization of broiler and layer birds and found one putative region including TSHR that was associated with metabolic regulation and photoperiod control of reproduction in vertebrates. The TSHR selective sweep may represent a significant feature of domestic animals, i.e., the restriction of seasonal reproduction that is now absent from domestic animals. In broilers, the selective sweep regions contained the genes IGF1, PMCH1 and TBC1D1, which are related to growth, appetite and metabolic regulation.

In pigs, putative selective sweep regions on SSC1 and SSC3 have been observed (Groenen et al., 2010a). The regions containing the genes IGF2, PRLR and GHR also had undergone possible selection.

Genetic diversity and genetic relationship analyses : Population genetics studies of domestic animals focus on genetic variability within breeds and genetic distances between breeds. Their purposes are to unravel the possible historic events during domestication and breed formation and assist in preserving the genetic diversity within endangered indigenous breeds. These studies will be helpful for scientific conservation and preservation measures, and for clarifying the population stratification for genomic selection and GWAS.

Genetic relationships among 19 cattle breeds with different geographical distributions were analysed using 37,470 SNPs during the Bovine HapMap Project. When the population was divided into two groups (K = 2) using Bayesian approaches, the cattle from the taurine and indicine breeds could be distinguished and crossbred populations showed admixture characteristics. Assuming nine groups (K = 9), most of the analysed cattle breeds could be classified into separate groups. Recently the phylogenetic relationships among 372 animals from 48 cattle breeds were characterized using the BovineSNP50 array. The results were consistent with the biogeography of breeds but also clearly depicted the admixed nature of many populations and revealed pedigree relationships between individuals (Decker et al., 2009).

Kijas et al. (2009) analysed the genetic relationships among 403 individuals from 23 sheep breeds and 210 individuals from two wild sheep species with 1,536 SNPs. The genetic variability within both African and Asian sheep breeds were lower than those of European breeds, and genetic distances between individuals from African and Asian breeds were smaller than those of European breeds. The genetic relationships among breeds were consistent with the geographical distribution and history of breed formation. Close phylogeographical structure, high genetic similarity and low differentiation were observed in sheep breeds, which was in agreement with the previous findings from other genetic markers.

vonHoldt et al. (2010) detected genetic relationships of 912 dogs from 85 breeds and 225 grey wolves using 48,000 SNPs. Both the neighbor-joining (NJ) clustering tree based on SNP genotypes of individuals and population clustering based on haplotype similarity showed single breeds could be distinguished from one another and grouped into Asian, Middle Eastern and Northern groups, which were consistent with the history of breed formation. In addition, domestic dogs had a higher proportion of multi-locus haplotypes unique to Middle Eastern grey wolves, suggesting that domestic dogs may originate from the Middle East instead of the Far-east as previously hypothesized.

Breed clustering is not always as successful as it was in the previous studies. Wade et al. (2009) analysed the genetic relationships between 11 horse breeds using 1,007 SNPs and found that the relationships between the studied breeds could not be clarified. This result may be due to the close relationships among domestic horse breeds.

Muir et al. (2008) examined genetic variability of chickens representing commercial, experimental and standard breeds using 2,551 SNPs. Based on the proportion of missing alleles and inbreeding coefficients, commercial broiler and layer line birds were found to have lost a significant amount of genetic diversity (~50% or more) from ancestral breeds. It was suggested that genetic diversity could be recovered within lines by crossing multiple pure lines from chicken breeding companies.

The high-density SNP array has also been useful in understanding the phylogenetic relationships of domestic animals. The dog was determined to be most closely related to the grey wolf, followed by the coyote, the golden jackal and the Ethiopian wolf (Lindblad-Toh et al., 2005). For pecoran (higher ruminant) species, 17 novel relationships were identified and another 16 previously proposed nodes within the infraorder were confirmed (Decker et al., 2009).

CNV Detection

Copy number variation (CNV) refers to a DNA segment that is 1 kb or larger and has variable numbers of copies in comparison with a reference genome. CNVs generally occur in more than 1% of the population, and they have often been found to be associated with specific diseases in humans. The comparison of the fluorescent signal intensity ratios of alleles at each SNP across the genome based on the Illumina BeadChip platform is one approach for CNV identification (http://www.illumina.com). Matukumalli et al. (2009) predicted 79 CNVs in diverse cattle breeds using the BovineSNP50 array, and ten of them were verified by comparative genome hybridization (CGH) array genotyping results. Fan et al. (unpublished data) predicted 12 CNV regions in pigs with the PorcineSNP60 array and found two large CNV regions of interest on SSC14.

Other Applications

High-density SNP arrays have been used for relationship and paternity testing and tracing the geographic origins of animal products. The arrays were also utilized for constructing high-resolution linkage maps, improving physical mapping orders, exploring the potential relationships between genomic sequence features and recombination rates, and carrying out linkage disequilibrium and linkage analysis (LDLA) mapping in particularly interesting regions.

Future Prospects

Even though SNP arrays have been widely applied in animal breeding and genomics, they still have some limitations with regard to the coverage and annotation of probes on the arrays. The coverage of the arrays for certain species is still low and uneven. Some genomic regions have very few SNPs. Further population information from HapMap studies could potentially solve this issue in combination with new SNP arrays. In addition, on some of the currently released commercial SNP arrays, there are still a number of unmapped SNPs. For example, ~1,800 SNPs on the BovineSNP50 array were unassigned based on Btau4.0, and ~8,000 SNPs on the PorcineSNP60 array were unmapped based on Sus scrofa build 9. WGAS have uncovered some associations with unmapped SNPs, and a few of them could be localized by LD estimates with the mapped SNPs. Additionally, the physical locations of some mapped SNPs have been corrected with the production of new genome assemblies (Ramos et al., 2009). Therefore, it is necessary to continuously improve the genome assembly and assign these SNPs to the correct physical positions.

Another issue is the annotation of the findings from SNP arrays. According to the reported GWAS, most of the trait-associated SNPs (TASs) were located in genes without obvious biological significance on the analysed phenotypes, or they were located in the intergenic regions or introns of certain genes. Similarly, in GWAS in humans, the TASs were not always in or near putative candidates relevant to the diseases (Manolio et al., 2009). Furthermore, a statistical summary indicated that TASs were intronic (45%), intergenic (43%), exonic and nonsynonymous (9%), exonic and synonymous (2%), or in a 5' or 3' untranslated regulatory regions (2%) (Hindorff et al., 2009). These unexpected results may be due to i) the TASs may be from genes that have not yet been annotated or may be unmapped SNPs demonstrating that further annotations of the current genome assemblies are necessary; ii) the sample size (especially for lowly heritable traits) and the genetic backgrounds of the studied populations may have effects on the association analyses, and multiple pure breeds and larger samples sizes may be of help; iii) the large (~40 Mb) average interval length between SNPs and uneven SNP distributions of the current arrays are major limitations for haplotype block analyses and fine mapping, so higher density SNP panels may be worth developing for improved analyses depending on the extent of LD in the analysed populations; and iv) the robustness of statistical methods for GWAS could be improved.

As more and more studies using SNP arrays become available, effective storage of the original data and curation of results could be other important issues. With the emergence of the large WGAS and population genetics studies, it will be feasible to build databases related to genome variation and/or candidate genes as public repositories, facilitating the comparisons of data across studies. In humans, several genome variation and GWAS databases have been developed. Ogorevc et al. (2009) constructed a gene database on cattle milk production and mastitis traits, but the capabilities of this database are limited. The designed databases should be comprehensive toolkits, interactive with whole-genome sequence, QTL mapping results and as much other related information as possible (Hu Z-L, personal communication). Such comprehensive databases will contribute to better utilization by the community of researchers doing animal breeding and genetics research.

Lastly, statistical analysis of SNP array data still presents a challenge. The large volume of data generated by SNP arrays is computationally demanding, which requires more sophisticated statistical models and efficient analytical methods. Statistical methods for genomic selection and GWAS are always being developed and improved. The approaches derived from different theories and algorithms will certainly impact the accuracy of the analyses. In addition, most early genomic selection studies were performed with simulated data, which are quite different from real data that often have limited sample sizes, and which may lack detailed pedigree information. Therefore, novel efforts are still needed in quantitative genetics, population genetics, and bioinformatics to develop advanced and efficient statistical approaches that will improve the applications of high-density SNP arrays in animal scientific research and production.

Sperm-Mediated Exogenous DNA Transfer Method

Transgenic technology holds considerable promise to advance understanding in biomedical and agricultural systems with some believing that one day transgenic animals may directly contribute to farming and breeding practice. Practical applications of transgenics in livestock production include improved milk production and composition, increased growth rate and disease resistance, improved feed usage and carcass composition, enhanced reproductive performance and increased prolificacy. Recently, transgenic technology has also provided a method to transfer nutritionally beneficial traits to other foodstuffs (Arihara, 2006; Maga et al., 2006). The first transgenic livestock were born 20 years ago (Hammer et al., 1985) and there are numerous potential transgenic methodologies to generate transgenic animals. Today, the most widely used methods for the production of transgenic farm animals are direct microinjection of foreign DNA into the pronuclei of fertilized eggs, nuclear transfer using genetically modified embryonic or somatic donor cells and viral-based constructs as vectors for the introduction of exogenous DNA into embryos. These methods have been restricted in part by inefficiency when applied in livestock. Moreover, it must be noted that the use of retroviral vectors is affected by safety issues.

The spontaneous capability of sperm cells to bind exogenous DNA molecules and internalize them into nuclei can be exploited by using spermatozoa as vectors for delivering foreign genetic information to eggs during fertilization. Sperm-mediated gene transfer (SMGT) appears to be a simple, efficient and relatively inexpensive method for modifying animals and their genome. It is also potentially applicable to every species using spermatozoa for its own propagation. Lavitrano et al. first described the SMGT procedure in mouse with high efficiency. Subsequently, they successfully adapted and optimized the technique for use in large animals. It was highly efficient for the generation of human decay accelerating factor (hDAF) transgenic pig lines. Several studies have proven the broad applicability of this technique to different animal species, from sea urchin to cattle. DNA uptake by mammalian spermatozoa is a very specific and highly regulated phenomenon. A considerable number of studies performed since then have revealed some unexpected aspects of spermatozoa, which suggest that such an attractive route to transgenesis may not be so easy but could be still be available if only the underlying mechanisms were understood.

Because of the outstanding protein synthetic capacity of the mammary gland and appropriate body size, the goat (Capra hircus) is the ideal animal for transgenics to produce recombinant proteins in a more efficient manner than other livestock. However, at present there are few reports of transgenic goats by SMGT. This study was carried out in order to establish technical protocols for SMGT for goats and evaluate the possibility of producing transgenic goats by the method of artificial insemination (AI) using sperm cells to integrate the exogenous DNA.

Materials and Methods

Experimental Animals and Management

This study was conducted at Rongchang goat farm (in western Chongqing City), Xichong goat farm (in northern Sichuan province). The bucks (sperm donors) were housed at the facilities of The Third

Military Medicine University. The bucks were Boer, Yudong White goat and Nanjiang Yellow goat. The doe breeds were Guanzhong Dairy goat and Nanjiang Yellow goat. The does that had kidded at least once, aged between 2 and 6 years and weighing 28-40 kg, were group-housed under natural photoperiod. The does were allowed access to an outdoor concrete run and fed a daily ration of 200-300 g concentrate. In addition, all experimental goats were allowed to an unrestricted access to straw, mineral salt lick and water.

Experimental Design

In order to set up and compare the technique with conventional processed semen, and to evaluate the effect of the reduction of spermatozoa number on fertilization rate, a total of 134 estrus induced does were divided into 4 groups randomly.

Plasmid preparation and DIG DNA labelling

The plasmid pEGFP-N1 encoding green fluorescent protein reporter gene was prepared at a large scale as described by Sambrook and Russell (2001), and extracted by plasmid maxi kit (Omega Biotek, Norcross, GA, USA). The extracted plasmid was digested by ApaII or NotI (Roche, Basel, Switzerland), and resuspended in filtered sterilized water.

Some plasmid DNA was digested by Hind III (Roche) and Dig Labelled. The DIG DNA Labelling was according to the kit protocol (DIG DNA Labelling and Detection Kit, Roche). Briefly, lineared plasmid DNA was diluted in autoclaved, double distilled water to a total volume of 15 and denatured by heating for 10 min in a boiling water bath and quickly chilled on ice before adding 2 [micro]l hexanucleotide mix, 2 [micro]l dNTP mixture and 1 [micro]l Klenow enzyme. This was mixed and centrifuged briefly, then incubated at 37[degrees]C overnight. The reaction was stopped by adding 2 [micro]l EDTA (0.2 mol/L, pH 8.0). The labelling efficiency was quantified with a dot blotting method on one strip of nylon membrane. A dilution series of the labelled DNA and the control DNA were prepared as described in the kit.

Semen Collection and Processing

Semen samples were collected with a pre-warmed artificial vagina. The treatment and washing procedure was as described previously (Zhao et al., 2005). The ejaculates were immediately diluted 1:4(v/v) in pre-warmed TALP buffer and washed (500 g for 5 min) at room temperature. Then, one aliquot of each ejaculate was immediately transported to the laboratory. Semen cryopreservation was performed as described (Zhao et al., 2008). For cryopreservation, briefly, the spermatozoa were immediately extended with Tris buffer (Tris 4.039%, citric acid 2.316%, glucose 0.667%, egg yolk 20%, glycerol 5.3%, penicillin 100,000 IU/100 ml, streptomycin 100,000 IU/100 ml) at a concentration of 200×106 spermatozoa. After that, the samples were maintained at 2-5[degrees]C for 4 h. Finally, the aliquot was cryopreserved as 0.1 ml pellets and stored in liquid nitrogen (-196[degrees]C) for at least 15 days. The pellets were thawed in TALP buffer at 40[degrees]C for 1-2 min.

Goat sperm were incubated with DNA in vitro as described in (Zhao et al., 2005). In short, semen (fresh or frozen-thawed) was immediately diluted in pre-warmed TALP buffer and washed three times at room temperature. About 106 sperm cells were incubated with 1,000 ng plasmid DNA at 20[degrees]C in TALP buffer for 60 min.

The DNA-treated sperm were washed three times in 0.9% NaCl and then fixed in 4% paraformaldehyde in PBS for 30 min. The sperm were smeared on gelatin-coated slides, dried and fixed. The sperm slides were rinsed in 0.01 mol/L phosphate-buffered saline (PBS) twice then in Washing buffer (Tris-HCl 100 mmol/L, NaCl 150 mmol/L, pH 7.5) for 10 min.

They were then incubated in Blocking solution (Dissolve Blocking reagent 10% (w/v) in washing buffer while constantly stirring on a heating block (65[degrees]C) or heat in a microwave oven, then autoclave. Prepare a 1xworking solution by diluting 10xBlocking solution 1:10 with Washing buffer) for 30 min. Anti-Digoxigenin-AP was prepared by centrifuging for 5 min at 10,000 g, then pipetting the necessary amount carefully from the surface and diluting it 1:1,000 in Blocking solution. Binding to the DIG-labelled probe in Antibody solution was performed for 30 min, followed by equilibration for 5 min in Detection buffer (Tris-HCl 100 mmol/L, NaCl 100 mmol/L, pH 9.5) and incubation in 2 ml freshly prepared Colorsubstrate solution (Add 200 [micro]l of NBT/BCIP to 10 ml of Detection buffer) for 60 min in the dark. The reaction was stopped in TE-buffer (Tris-HCl 10 mmol/L, EDTA 1 mmol/L, pH 8.0) and the slides washed for 5 min with 50 ml double distilled water. Finally, slides were examined and photographed under a microscope (Olympus CX41) at 1,000x magnification.

Synchronized Estrus Induction

The does were treated using intravaginal sponges impregnated with 30 mg Levonorgestrel (Levonorgestrel, Beijing Zizhu Co., LTD) for 12 days, then intramuscular injection of 25 IU FSH (FSH, Ningbo Hormone Co., LTD) and 0.05 mg [PGF.sub.2[alpha]] ([PGF.sub.2][alpha]], Ningbo Hormone Co., LTD) at sponge withdrawal. All does were checked once daily to ensure that sponges remained in place during the treatment period (Baldassarre and Karatzas, 2004). Does that had lost their sponge were noted and the sponge was immediately replaced. Estrus was detected at 6 h intervals by visual observation (06:00, 12:00 and 18:00 h) during the 5 days following the end of treatment, using two intact teaser bucks fitted with an abdominal apron. Does were considered to be in estrus when standing to be mounted.

Intra-Cervical Insemination

Intra-cervical inseminations were performed by lifting the hindquarters of the doe over an elevated rail while the front legs remained standing on the ground. Each doe was inseminated within 4 h after semen preparation. The semen was deposited into the external opening of the first cervical fold, using a speculum inseminating pipette according to standard procedures (Salvador et al., 2005).

Intrauterine Insemination

Intrauterine inseminations were performed on animals under local anaesthesia by injection of 4 ml procaine hydrochloride. Does were placed head down and supine on a 30[degrees] sloped surgical table with the hind legs tightly bound to the table in order to support most of the body weight after preparing for surgery. Intrauterine inseminations were performed as described. Briefly, after surgical preparation, the presence of mature preovulatory follicles was checked and the insemination needle with semen was inserted through the uterine horn. An extra volume of 0.5 ml of extender was used to wash out all the spermatozoa from the needle. Antibiotic cover was given by Tetracycline (20 mg/kg). Recovery was usually complete by 2-4 days post-treatment. Conception rates were determined by 25-day non-return rates.

PCR and Southern Blotting Analysis

The genomic DNA from the kids was detected by screening using PCR technology with primers: forward primer

(5 '-CCCTTCCCTCGTCTCCAC-3') and reverse primer (5'-CGTCGCCGATGGGTGGTGTTCT-3'). Each PCR sample consisted of 25 [micro]l of solution containing 10-20 ng of template, 2.5 [micro]l 10xBuffer, 2.5 [micro]l MgCl2, 2 [micro]l dNTP, 0.5 [micro]l of each primer, 0.5 Taq polymerase (TaKaRa). PCR consisted of 35 cycles of denaturation at 94[degrees]C for 45 s, annealing at 60[degrees]C for 45 s, and extension at 72[degrees]C for 75 s, followed by 10 min extension at 72[degrees]C. Each PCR sample was subjected to electrophoresis on a 1% agarose gel.

After digestion with AseI and NotI completely, genomic DNA was separated electrophorectically on a 0.8% agarose gel and transferred to a nitrocellulose membrane. After air-drying, the DNA was cross-linked to the membrane by UV irradiation, and then hybridized with a biotinylated probe for the pEGFP-[N.sub.1] gene (Detector[TM] PCR DNA Biotinylation Kit, KPL). Hybridization and washing procedures were as described previously (Zhao et al., 2008). Hybridization was carried out overnight at 42[degrees]C in a standard buffer solution (DNA Detector[TM] HRP chemiluminescent Blotting Kit, KPL) containing 50 ng/ml of denatured probe. The washing and detection procedures were the same as described by Sambrook and Russell (2001).

Statistical Methods

The mathematical model for kidding percentage was compared by one-way analysis of variance. Differences between means were tested for significance by the Duncan's new multiple range test (Weiss and Hassett, 1984). PCR and Southern blotting positive rates were analysed using Yate's corrected [chi square]-test of independence and $p < 0.05$ was considered statistically significant (Snedecor and Cohran, 1980). The seminal parameters of bucks were compared by t-test. The statistical software program SPSS 13.0 was used for all analyses.

Results

Sperm Quality and Transfection Efficiency

We selected two bucks, one Boer goat (B2029) and the other a Yudong White goat (YW011), as the sperm donors. The seminal parameters of the selected bucks. The differences in ejaculate volume, motility, acrosomal integrity and sperm concentration between the two bucks were not significant ($p > 0.05$). However, the abnormality rate of B2029 was significantly higher than that of YW011 (t = 2.772, p = 0.024) and the live index was more significantly lower (t = 6.299, p = 0.003). Both sperm donors had a high ability to pick up exogenous DNA and no significant difference was found (64.09% [+ or-] 4.56% and 55.67%+1.77%, t = 1.5118, p = 0.1910).

Kidding Results

The kidding results for different insemination deposition sites and sperm numbers. A total of 57 does kidded. There were 30, 20, 2 and 15 does kidded of Group I to Group IV, respectively. The kidding percentage was very significantly different among groups ([F.sub.3, 130] = 5.501, p = 0.001) and it varied between 12.50% and 62.50%.

Transgenic Kids Produced by Sperm Mediation

A total of 76 kids were generated. The presence of the pEGFP-[N.sub.1] DNA was screened by PCR and also by Southern blotting analysis.The results of different groups. Eight kids were tested positive on the basis of agarose gel electrophoresis of the PCR-amplified fragment. Southern blotting analysis of the samples showed 5 positive kids.

Discussion

We describe herein the possibility of SMGT producing transgenic goats. The results showed that transgenic goats could be produced efficiently with the method of artificial insemination by the sperm cell integration of the exogenous DNA. A total of 76 kids were generated, among them 5 kids tested positive by Southern blotting analysis. The major benefits of SMGT were found to be high efficiency, low cost and ease of use compared to other methods. Traditionally, the production of transgenic animals has relied almost exclusively on the microinjection technique. The technique was first developed in mice, and the first batch of genetically engineered supermice was generated by applying the same technique. However, that procedure is relatively complex, inefficient (with an efficiency of about 3%), and expensive. When applied to other species these methods either work with lower efficiency or do not work at all. SMGT shows high efficiency in mice. Subsequently, this technique was successfully adapted and optimized for use in large animals with high efficiency.

The key feature of this study is that we have obtained a number of transgenic goats with an efficiency of 6.58% by SMGT method. The selection of sperm donor also plays an important role in producing transgenic goats. Lavitrano demonstrated that there are two important parameters that must be optimal for the SMGT technique to be effective in swine: selection of sperm donor and optimization of DNA uptake. In our former study, we have found that different goat sperm donors appeared to differ in ability to bind foreign genes (Zhao et al., 2005). The standard sperm parameters, such as volume, concentration, presence of abnormal sperm, motility and the ability to bind foreign genes, were assessed one by one. We then selected two bucks, one Boer goat and another Yudong White goat, as sperm donors. We concluded that these two bucks could supply high quality semen and serve as good vectors for exogenous DNA.

Our study showed that frozen-thawed spermatozoa could also be used as a vector for introduction of exogenous DNA and produce transgenic goats. This is the first report of the SMGT method in this field. Anzar and Buhr (2006) found transfection efficiency of bull spermatozoa to be greater in frozen-thawed than in fresh spermatozoa, both from the same ejaculate. We have investigated the efficiency and the ability of fresh and frozen-thawed sperm to pick up exogenous DNA, and found that frozen-thawed goat sperm were more efficient and more reliable than fresh sperm cells (Zhao et al., 2008). The transgenic kid ratio was lower than the transgenic embryonic ratio, but freeze thawing treatment could enhance the efficiency of picking up exogenous DNA and producing transgenic kids. Due to low temperature and high salt concentration, semen cryopreservation reduced the integrity of sperm plasma membranes and enhanced sperm genomic DNA strand breakage significantly.

The data presented in this work demonstrate that intrauterine insemination may be suitable for reducing the number of frozen-thawed spermatozoa to one million and a gain in kidding percentage of 36.59%. Large doses of fresh semen are used in conventional artificial insemination. However, the number of spermatozoa can be greatly reduced if they are deposited close to the site of fertilization, thus diminishing their loss along the uterine horns. In this work, the production of transgenic kids by SMGT is based upon the spontaneous uptake of exogenous DNA by goat spermatozoa removed from seminal plasma. A single standard dose for goat artificial insemination (AI) requires a huge amount of DNA which is expensive and time-consuming to obtain and purify. Also, the experimental bucks were housed in the University facilities, whereas the does were at goat farms far from the University. The long-term frozen storage of semen was very necessary for performing artificial insemination efficiently. In order to achieve a high kidding percentage with low numbers of frozen sperm, intrauterine deposition of semen is required. However, the kidding percentage in our study was not satisfactory. There are a number of possible reasons. The first is surgical laparotomy that results in distress for the does and in postoperative adhesions complicating either subsequent embryo production or delivery. The second is the low number of spermatozoa. The total number of spermatozoa inseminated per female is one of the main factors affecting fertility. It was found that 200 million frozen-thawed spermatozoa was sufficient to obtain acceptable kidding performance in conventional artificial insemination. Intrauterine insemination resulted in improved fertility rates and allowed a reduction in numbers of frozen-thawed spermatozoa inseminated to 5-20 million motile spermatozoa (Ritar et al., 1990). One million sperm could decrease significantly the amount of exogenous DNA required, but result in a low kidding percentage. The third is that the does were induced to estrus. Natural, synchronized or induced estrus can influence fertility. In our study, the does were synchronized. Despite improvements in the efficiency of hormonal synchronization of estrus, variability in the time of ovulation after treatment is an important impediment to attaining acceptable fertility after only one insemination with low number sperm. Goat intrauterine insemination with frozen goat semen should be optimized in the future.

Historical Geography of Estonian Cattle Breeds

The geographical distribution of cattle breeds is a topic not given much attention by the international research community. While there are some general studies and encyclopaedias that provide general data about the distribution of breeds, no real detailed research exists. However, during the last decade, much work has been done in the evaluation of the genetic material of cattle herds and the resultant variations within them. Studying the genetic material of cattle undoubtedly adds a new perspective to the history of European agriculture and sheds light on the geographical distribution of cattle breeds. This article, utilising the methods of historical geography, studies and analyses data from different cattle pedigree books, information about the distribution of cattle, and facts about the historical conditions of cattle breeding in Estonia.

The territory of the present-day Republic of Estonia has been subjected to occupations of several foreign powers throughout its history. Estonians have been ruled by the Danes, the Poles and the Russians; its territory has been divided, sometimes even simultaneously, between the aforementioned nations into several parts. The longest rulers though were the Germans, who became local aristocracy over the centuries, even during the reign of other nations over Estonia. This complicated political history has left its mark on the development of administrative areas within the country. Estonia was relatively homogeneous in its population and landscape until its conversion to Christianity at the beginning of the 13th century, at which point it was divided into North-Estonia, which belonged to the Danes, and South-Estonia, that belonged to the Livonian Order (branch of the Teutonic Order) and the Bishop of Riga. This border between North and South was determined by the natural geographical characteristics of Estonian territory—specifically bogs. This border area was, and still is, traversed by an East-West directional zone of bogs (Kulvik et al. 2000). This eventually became a permanent administrative border as knighthoods developed in the region. The German-speaking aristocracy formed one knighthood in North-Estonia (Estonia Province, capital in Tallinn), while South-Estonia together with North-Latvia (Livonia Province, capital in Riga) had their own knighthood. Additionally, Saaremaa had its own knighthood separate from the others. The governors of these separate provinces had their own privileges, laws, currency and units of measure. Although Estonia became a part of Tsarist a Russia after the Great Northern War (1700-21), its tight connections with Western Europe, the original founders of the northern and southern provinces, were sustained. While formally belonging to the Russian Empire, the Baltic Provinces enjoyed the Baltic Landesstaat (1) and the large population of Baltic Germans that lived in the region kept close contacts with their native land. These close connections facilitated the rapid transfer of cultural and economic developments that were taking place in Western European countries at the time to the eastern coast of the Baltic Sea. Compared to other regions of the Russian Empire, the Baltic Provinces were more advanced in terms of economic and cultural development, including cattle breeding. (2) This historical border and connections with Western Europe would eventually play an important part in the development of cattle breeding in Estonia.

Historical Development of Estonian Cattle Breeds, Historiography

While the history of cattle breeding in Estonia has been examined by researchers, the geographical distribution of cattle breeds has been of little interest. Estonia has been an agricultural country throughout its entire history. At the beginning of the Common Era, agriculture became a significant source of subsistence. By the end of the 19th century, it had become the most important source of industry and remained the leading industry within the Republic of Estonia during the 1930s.

The first written records of cattle breeding in Estonian territory originate from the Livonian Chronicle written by Henry of Livonia at the beginning of the 13th century (The chronicle of Henry of Livonia 1961). The Chronicle indicates that Estonians already had many cattle at that time and that from a campaign to Soontagana in 1210/11, four thousand cattle were taken as plunder.[3] The sizes of herds were also praised by other chroniclers, like the Russow Chronicle from the 16th century (The Chronicle of Balthasar Russow 1988). The first known data that confirms the search for a cattle breed suitable for the territory of Estonia dates all the way back to the 17th century. There are records indicating that Dutch cattle were bred in Purtse manor, Viru County, Estonia in 1624 (Karelson 1981). Furthermore, in the first thorough description of the territory of Estonia—"Topographische Nachrichten von Lief-and Ehstland" by August Wilhelm Hupel—the author mentions that cattle were bred quite extensively in Estonia and that the animals were smaller than those in other Russian provinces, Germany, and Denmark (Hupel 1777). Unfortunately, the text did not mention the breeds of cattle. Individual reports on the breeding of different cattle breeds in the manors of Estonia originate from the 18th century and from the beginning of the 19th century. According to Meinhard Karelson (1981), an expert on the history of Estonian cattle breeding, almost all popular Western European cattle breeds existed in Estonia in the 1850s. However, unlike in other European countries, no systematic breeding work with cattle was practised. Many cattle breeds were not suitable for the climatic conditions of Estonia and Livonia, as they originated from areas with different climatic and nutritional conditions. Consequently, the oldest agricultural society in the Baltic Provinces—the Livonian Benefit and Economic Society (established in 1792)—failed in improving the Estonian native cattle with Arshire oxen because the climatic conditions of Estonia were unsuitable for cattle from the Scottish highlands (Lepajoe and Oll 1998).

Cattle breeding eventually became more systematic in the Baltic Provinces under the Russian Empire in the second half of the 19th century, mainly because of increasing competition in the European market of agricultural products. In order to ensure a market for their agricultural production, the Baltic German landlords had no other choice than to increase production and find ways to improve the native cattle. Alexander Theodor von Middendorff (1815-1894), a Baltic German scientist, was the most notable cattle breeder during this period in Livonia.

His Baltic German background was somewhat interesting as his father was a nobleman, while his mother was an Estonian peasant. Middendorff was born in St. Petersburg and attended school there at an early age, but later moved to Tartu and attended Tartu University to study natural sciences under the Faculty of Medicine. Since his childhood, Middendorff had been an enthusiastic hunter and always very interested in rural life, as he was very active in the management of his father's estate in Livonia. During an expedition to northern and eastern Siberia, which lasted from 1842 to 1845, he made significant observations on agricultural practices and taxation systems employed in the regions.

The expedition expanded his scientific reputation, but had negative effects on his health and physical condition. Middendorff eventually bought his own estate in Livonia in 1850 and continued to pursue his interest in the complex issues of Livonian agricultural policy, the regulations of which were mainly in favour of the interests of Baltic German landlords. A reform plan launched in 1840 suggested a shift from corvee to personal freedom and the selling of land to peasants; however, it was not binding for the landlords to follow this plan. Middendorff, who was a member of the St. Petersburg Academy of Sciences and an elected member of the Academy of Zoology, travelled extensively in Russia and Western Europe, and as a result, obtained extensive knowledge on the overall practices in agricultural policy. He observed that various European countries gradually introduced common agricultural regulations similar to each other, which leaned towards large farms with intensive production.

Russia, including Livonia, still followed old agricultural practices. Middendorff realised that Livonia might lose its agricultural market and suggested that new scientific methods should be applied, both by landlords and peasants, in order to compete with other European and Ukrainian producers, who lived in more favourable climates. In 1861, Middendorff returned to Livonia from St. Petersburg due to health problems and became head of the Livonian Benefit and Economic

Society (1862-1882). Its membership was not accessible to everyone, it had little authority to introduce new practices to the public. Middendorff made an effort to open the Society to all interested parties with the intention of providing everyone with information about new methods in agricultural production. This would enable Livonia, as a whole, to become agriculturally more fruitful. The key issue to improving Livonia's agricultural development, Middendorff believed, was the improvement of dairy production. He considered the raising of dairy cattle to be more favourable than beef cattle (Tammiksaar 2006, Sukhova and Tammiksaar 2005).

Even though many of the leading Livonian manor farms were qualitatively, as well as quantitatively, productive, Middendorff claimed their outputs were not enough to guarantee the competitiveness of Livonia's agricultural products. He once stated:

Although he claimed that dairy farming was the basis for a boom in agriculture, Middendorff considered the intensification of agricultural production in all strata to be important for the development of Livonia. However, development would not flourish without systematic breeding work with native cattle, as cattle breeding was the way to stay competitive in the dairy market. To improve the pedigree of native Estonian cattle, a suitable breed accustomed to similar climatic conditions had to be found. In the late summer of 1862, Middendorff found several candidates from among the Red Angler cattle breed in Schleswig-Holstein and Denmark and chose 31 Anglers from among 880 in the area of Tondern and the Lugum Monastery in Denmark to begin breeding. (5) With these animals, his aim was to breed cattle that could produce a high milk yield and could adapt to living in the natural conditions of Livonia. By 1863, he had achieved very positive results with the Angler breed and the new cattle became quite popular among the landlords of Livonia. The import of a new cattle breed and the uncontrolled improvements of the local breed led Middendorff to begin identifying the characteristics of the native cattle breed in order to control the improvements. To do this, accurate data on the origin, age, calving, milk yield, milk quality, and provender of the cattle had to be gathered (Baltische... 1871). Landlords and peasants followed Middendorff's data collection instructions and by 1879, Livonia had achieved notable advancements in improving the breed (Baltische... 1879). The next step in raising the breed value of the new cattle was recording the individual breed characteristics of the best specimens into pedigree books. In 1884,

Middendorff proposed to start recording the breed characteristics and in 1885 the Baltic Cattle Breeders Association was founded, which kept strict records of different cattle breeds in pedigree books (Baltisches..., 1886). The Breeders Association decided that the weight of the new developing breed needed to be increased and so the import of larger Danish Red Funen cattle to Livonia began in 1892 (Magi 1924). The mixing of the two imported Danish breeds and the native cattle served as a basis for the creation of a new Livonian Red cattle breed and, subsequently, was responsible for a dairy farming boom in Livonia. By the end of the 19th century, the Province of Livonia had become the most advanced agricultural area in the entire Russian Empire.

North-Estonia (the Province of Estonia) began developing its own breed of cattle in the middle of the 1880s, considerably later than the cross-breeding efforts in South-Estonia (the Province of Livonia). Whereas Angler and Funen breeds were brought to Livonia, the Holstein and the East-Friesland cattle breeds, which also produced high milk yields, were imported to Estonia (Bodisco 1894). The instigator of agricultural progress in Estonia was the Estonian Agricultural Association, which was founded in 1844. The association's first real leader was Alexander von Keyserling (1815-1891), a palaeontologist from Kurland, who took control of the organization in the 1850s. He originally studied at Berlin University and was later employed in Russia as a civil servant. In the early 1840s, he organized three expeditions into the European part of the Russian Empire (two of them with Sir Roderick Murchison). Keyserling's academic career did not last long as his wife frequently suffered from poor health and his family decided to move to Estonia where his father-in-law, Feodor Cancrin—the Minister of Financial Affairs of Russia, had given them several manor estates as a wedding present. Keyserling was very active with the agricultural success of his estates and successfully intertwined them with the activities of the Estonian Agricultural Association.

Count Leo von Keyserling (1849-1895), the son of Alexander von Keyserling, had even more influence in the promotion of agricultural policy within the Estonian Agricultural Association. Having a wide-ranging knowledge of agricultural practices, he continued his father's responsibilities both in management of his estates and leading the Association in 1893. Although both father and son had much respect for Middendorff, they believed the introduction of the Holstein cattle to the region would bring about better economic prosperity as the Holstein produced a higher milk yield than Middendorff's Red cattle. Count Keyserling dedicated all his energy to improving the breed of Estonian dairy cattle (Kivimae 1994). As a result of cross-breeding the Holstein and the East-Friesland cattle with the native cattle, the Estonian Black-and-White (also known as Estonian Holstein) cattle breed was created. Just like the Red cattle breed of Livonia, the characteristics of the Estonian Black-and-White cattle were recorded in pedigree books.

By the end of the 19th century, most Estonian (the Estonian Province and Northern part of Livonian Province) households were still raising native cattle. However, due to rapid economic development within the Russian Empire, by the beginning of the 20th century, the new distinct cattle breeds had replaced the native cows and had become prevalent in both provinces (Lepajoe and Oll 1998). After Estonia gained its independence in 1918, cattle breeding activities continued to flourish and dairy products became the country's most important items of export (primarily to Denmark and England). The Breeders Associations of Estonian Angler and the Breeders Association of Estonian Black-and-White Dairy Cattle were both established in 1919, and a Breeders Association of Estonian Native Cattle was established in 1920. Whereas breeding activities during the previous years were mainly the responsibility of individual landlords, these new associations worked with each other to improve cattle breeding within the entire country. A network of test stations was developed during the 1920s and 1930s, which organized breed shows and conducted artificial inseminations on cattle in order to help the breeds proliferate (Saveli 2006). These efforts ensured the stable development of all three breeds of dairy cattle in Estonia.

During the Soviet period, animal breeding was reorganized in Estonia with the creation of the Research Institute of Animal Breeding and Veterinary Medicine at the Academy of Sciences of the Estonian Soviet Socialist Republic in 1947. The work of the institute was controlled by the government, mainly based on the guidelines set forth in the "Means for developing agriculture in the post-war period" from the Central Committees of Estonian Communist Party XVI plenary session (5th of April 1947). According to this decree, the improvement in the quality of the breed of farm animals was of the utmost importance and should be an actively targeted goal. Organizations coordinating the breeding of cattle were founded and work with the Estonian Red cattle began in 1947 and with the Estonian Holstein cattle in 1948 (Polluaar 2004, Bulitko et al. 2004). These organizations became centres for cattle breeding; however, the main difference from the pre-war period was that now breeding was done mainly in collective farms. The organization promoting native cattle was not re-established in the Soviet period and thus might be the reason why the number of native cattle decreased drastically after the Second World War. In 1956, agricultural institutes were placed under the supervision of the Ministry of Agriculture in each of the Soviet republics, including Estonia. That same year, the Estonian Livestock Breeding and Veterinary Institute (ELVI) was established, which became the main centre for animal breeding until Estonia regained its independence in 1991, at which point the ELVI was joined with the Estonian Agricultural University (now Estonian University of Life Sciences). After the second independence in 1991, cattle breeding continued to take place in breeding organizations, except this time, the Estonian Native Cattle Association was re-established to coordinate the breeding of native cattle. Today, Estonian native cattle are on the endangered breeds list of the FAO (Food and Agriculture Organization) (Kalamees 2004, Kalamees 2007). In addition to the mentioned breeding organizations, an Estonian Beef Breeders Association was established in 2003 and all of these organizations were united under the umbrella of the Animal Breeders Association of Estonia in 1993.

Data and Methodology

This article analyses the geographical and numerical distributions of Estonian cattle breeds and the conditions that have caused the particular distributions from the 1860s until today. Moreover, we wish to check the claims of previous researchers that the Estonian Red cattle is bred mainly in South-Estonia and the Estonian Holstein cattle mainly in North-Estonia. Early statistical data used in this research paper has been obtained from an article on the development of dairy farming before 1918 by Kivimae (1994) and from books on agricultural statistics until 1940. The sources of data from the Soviet period are productivity control reports of cattle in 1976 (Idarand 1978). The most recent data come from the database of Estonian cattle breeds by the Estonian Agricultural Registers and Information Board (ARIB) from November 2005. The data from 1976 has been processed to be comparable with the contemporary administrative system and, because of this, some imprecision may occur on the distribution map of the cattle. However, this does not affect the overall picture of distribution. The distribution map of the cattle from 1976 does not cover the cattle bred in households; it covers only the cattle bred in collective farms, on which only the official data is available. Information about the borders between the Soviet rural municipalities originates from a CD of Estonian Geography (Roosaare et al. 2000).

The Number and Location of Estonian Cattle Breeds—Analysis and Results

As it appears, the ratio between the Estonian Holstein and the Estonian Red--the two most widespread breeds—was relatively equal throughout the territory of Estonia at the end of the 19th century. However, differences between cattle breeds due to location are notable. It has to be mentioned that most of the cattle in Estonia were not necessarily pedigree at the beginning of the 20th century. Even in the control year 1938/39, pedigree cattle formed only 11 percent of the total number of cattle. However, this percentage reflects only the number of pure pedigree cattle; the number of cattle having pedigree characteristics was certainly much larger. As the 20th century progressed, the ratio balance shifted in favour of the Estonian Red cattle breed. For example, in 1939, 73 percent of all cattle controlled were Estonian Red (Eesti karjakontoll... 1940); the reason for this being that South-Estonia was the most prosperous part of Estonia in those years (Kant 1935) and the cattle bred there were set as a good model for what cattle should be bred in other regions. Thus, the Estonian Red cattle breed expanded its area of distribution during the first half of the 20th century.

After the Second World War, the Estonian Red breed started to lose its superiority in distribution to the Estonian Black-and-White breed (Holstein) and after 1970, the ratio between the two breeds changed dramatically. The main reason for the change is that the Estonian Holstein breed is more productive as a dairy cow (6) and that Estonian cattle breeding is predominantly dairy farming oriented. The analysis of the protocols of the Academy of Sciences from the Second World War on the development of agriculture in Estonia during the second half of 1940s shows that a significant amount of attention was paid to the breeding of pure-bred cattle, but no preference in breed was made. The protocols set forth that the main purpose of breeding was to increase the milk yield and the amount of fat in milk, to improve the meat output, and to breed the three most common cattle breeds in Estonia, including the native cattle. The regionalization of breeds was also mentioned as it was planned to breed Estonian

Black-and-White cattle in the districts of industrial centres and holiday resorts and breed Estonian Red cattle in areas further from markets (Central and South-Estonia), where butter and cheese was produced. In more widespread areas, there existed a possibility to raise Estonian native cattle, which were ideal for producing milk with high fat content, in the form of micro-regions (Pung 1947). These protocols set a benchmark for the growth of Estonian Holstein cattle, as compared to Estonian Red cattle, simply because it was more productive as dairy cattle. This growth was quite constant without major fluctuations. As of 2005, 70 percent of cattle in Estonia were of the Estonian Holstein breed.

When analysing the distribution of cattle breeds in greater detail, it can be noted that the historical-provincial distribution of cattle breeds between North and South-Estonia still exists today. The only exception among the counties of the former territory of Livonia is that of Parnu County, where Estonian Holstein cattle were predominant until the end of the 20th century. In the other counties of South-Estonia, Red cattle were traditionally more common until the 1990s. There is no simple explanation for the popularity of the Estonian Holstein breed in Parnu County. Even though its location and historical roots to Livonia make Parnu County part of South-Estonia, its landscape, cultural, economic, and geographical characteristics, as well as dialect, are more similar to those of North-Estonia. This might account for why the development patterns of cattle breeds in Parnu County have been different from that of the rest of South-Estonia. Preference for the Estonian Holstein breed in Parnu County may also be connected to historical persons involved with cattle breeding in the area. When the Livonian Dutch Friesian Cattle Breeders Association was established in 1901, Otto Hoffmann, the landholder of Sauga Manor (near Parnu), was elected as the breeding inspector. Coincidentally, Hoffmann had worked for the Estonian Province Dutch Friesian Cattle Breeders Association; thus, the reason behind the popularity of the Black-and-White cattle in Parnu County, over the Red cattle, could be attributed to the personal influence of Hoffmann. Upon the analysis of the data in the first Livonian Pedigree Book of the Dutch Friesian Cattle Breed, it stands out that this breed was represented in all the biggest manors of Parnu County; in the manors of Taali, Sindi and Uulu (Stammbuch... 1902). Lastly, at the beginning of the 20th century, focusing on one certain cattle breed might have been set as an example by the bigger and more progressive farms, like the Piistaoja farm of the Pool family in Tori parish (Parnu County) where, again, Estonian Black-and-White were bred (Keevallik and Metsaalt 1981, Pool 1941). It is apparent that the favouring of the Estonian Holstein cattle breed in Parnu County (especially in the areas surrounding the town of Parnu) dates back to the very beginnings of cattle breeding there.

In North-Estonia, where mainly Estonian Holstein cattle are bred, the regions that differ from the norm are Ida-Viru County (eastern part of Estonian Province), Laane County and the Island of Hiiumaa (western part of Estonian Province). While the popularity of the Estonian Red cattle breed in Laane County and the Island of Hiiumaa can be explained by the compatibility of the cattle with the natural conditions there, their presence in Ida-Viru County remains unexplained.

Undoubtedly, the geographical distribution of cattle breeds in Estonia has been influenced by the so-called pedigree districts that were established in 1924 and divided the whole of Estonia into pedigree districts. Areas where several cattle breeds had been bred equally at the same time were named mixed districts. Prior to the establishment of these districts, farms that bred all three different steers could operate and receive state aid. Once districts were determined, farms were limited to only breeding specific breeds and exhibitions of young animals could be held only with approval of the district. If someone wanted to raise a different breed of cattle in a specific pedigree district, it had to be done at his own expense (Magiste 1939). According to a map of pedigree districts in 1939, the distribution area of the Estonian Red cattle breed covered the counties of Valga, Voru, Tartu, Viljandi and Petseri and partially covered the counties of Parnu, Jarva and Viru; the core areas of the Estonian Black-and-White where Harju and Jarva (Magiste 1939).

Estonia was also divided into breed regions during the Soviet period and these areas were decided by whether the area belonged to any industrial region or not. (7) The districts of the Estonian Black-and-White cattle breed in Soviet times were the counties of Harju and Rapla. The counties of Paide, Parnu and Rakvere were mixed districts and the rest of the counties were the districts of the Estonian Red cattle breed (Kutti et al. 1965). Estonian native cattle were not favoured during this period because its milk yield was insufficient and thus its number quickly decreased and it was usually replaced by the Estonian Red cattle breed that had managed to adjust to the natural conditions.

In general, the map of pedigree districts of 1939 agrees quite well with a map of pedigree districts of 1976. Thus, it can be concluded that preferring certain breeds in certain districts of Estonia remained almost unchanged for almost 40 years. Noticeable differences occurred only in Saaremaa and in West-Estonia. According to the map of pedigree districts of 1939, there were either only Estonian native cattle breed districts or mixed districts in West-Estonia. Nevertheless, by 1976, the Estonian Red cattle had gained a majority in West-Estonia and on the islands. Hence, it can be concluded that the Estonian Red cattle became dominant in West-Estonia and on the islands only after the Second World War.

After monitoring the changes in the geographical distribution of the cattle breeds over the last couple of decades, it becomes evident that the Estonian Holstein cattle breed has expanded its 'area of influence' to a great extent. Saaremaa seems to be the only county where the Estonian Red cattle breed is still predominant. This can be explained by the conservative tendencies of the islanders; many elements of Estonian folk culture have been best preserved on the islands too (Viires 2004). The natural conditions of Saaremaa could also be considered an important factor for the Red cattle's presence. Due to its thin soil, the ground of Saaremaa is quite infertile and lacks suitable meadows and grasslands, which would require breeding a relatively small-sized and less nourishment demanding cattle there. Even though Estonian native cattle originally dominated the area of Saaremaa, the Estonian Red cattle had moved in and taken command by the end of the 1930s (Pung 1947). As demonstrated in the data from 1976, by that time, Saaremaa had clearly become an area full of Estonian Red cattle.

In the traditional Estonian Red cattle breeding counties of Southeast-Estonia, the Estonian Holstein cattle breed has taken over. This is most noticeable around the town of Polva, where the biggest Holstein dairy cattle herds of South-Estonia are located (there are approximately 2,100 Holstein cattle in Polva Agro LLC). It appears that from among the ten counties with the highest number of cattle, only the counties of Poltsamaa and Vandra have a noteworthy number of the Estonian Red cattle breed.

From the data of 2005, we can draw various conclusions from the sizes of cattle herds. While the average size of an Estonian Holstein cattle herd is 22, the average number of an Estonian Red cattle herd is 11; half the size. Thus, it appears that Estonian Red cattle are bred in small-sized herds and farms. The size differences of cattle herds may be influenced by landscape conditions because the hilly landscapes of South-Estonia are usually more suitable for smaller herds, whereas the large plains of North-Estonia are more suitable for bigger herds. Most of the big herds in Estonia are of the Estonian Holstein breed (8) and are located, as mentioned before, in North-Estonia. During the last decade, the breed diversity of cattle has increased greatly. Traditionally there have been three main cattle breeds in Estonia (Black-and-White, Red, and native cattle), but nowadays, there are altogether 15 different cattle breeds represented in Estonia. These new breeds are mainly beef cattle. The first Hereford beef cattle were imported in 1978 (Suurmaa and Toi 2004) and have become the most numerous among beef cattle breeds in Estonia. It is an interesting fact that in three parishes (Korgessaare, Noarootsi and Taheva), there were more Hereford beef cattle than traditional Estonian native cattle breeds in 2005.

According to the distribution map of Estonian folk culture, Estonia is culturally divided into North-Estonia, South-Estonia, and West-Estonia together with its islands (Viires 2004). The traditional distribution of cattle breeds re-emphasises this cultural-geographical division within the country. Besides differences of folk culture and cattle breeds, there are other historical differences, mainly in church history, within Estonia that also originate from administrative divisions (Schmidt 1991, Pae 2006). Additionally, resembling the old administrative border, an East-West directional boundary of Silurian limestone and Devonian sandstone runs through Estonia (Mander and Palang 1994). Due to this geological border, traditional building materials of North-Estonia have always differed from those of South-Estonia. Likewise, regional dialects have separated North and South-Estonia for centuries. As we can see, the current research on geographical distribution of Estonian cattle breeds is just one more indicator of the cultural-geographical differences of North and South-Estonia

The following conclusions can be drawn from the current research on the geographical distribution of Estonian cattle breeds. As a result of the historical administrative division of Estonia between two provinces, the Estonian cattle breeds were geographically distributed into clearly defined districts until the 1900s. Systematic breeding work with cattle began in the second half of the 19th century and thus, responsibility for the coordination of cattle breeding was divided between provincial associations. Slight deviations from the provincially favoured cattle breeds occurred mainly in the counties of Ida-Viru and Parnu. Besides the border between the two provinces determining the geographical distribution of cattle breeds, their distribution can be attached to regional political, cultural, natural, personal and economic conditions as well. The current research also discovered that the numerical distribution of cattle breeds has changed greatly during the last half of the 20th century. While the Estonian Red cattle breed was dominant in Estonia until 1900, the Estonian Holstein breed eventually took over and has gained a majority due to its ability to produce a higher milk yield than other breeds. (9) The Estonian Red cattle's level of relative importance has fallen over time and it now only constitutes approximately 1/4 of the total cattle in Estonia. During the last 50 years, the number of Estonian native cattle, that has a low milk yield, has diminished to an extent where its survival has become questionable and it has been entered onto the endangered breeds list of FAO. Although cattle have been mainly used for dairy production, the number of beef cattle has increased in Estonia since its re-independence. As one can clearly see, the distribution, variety, and use of cattle breeds within Estonia has changed dramatically over the past several hundred years and these changes can be linked to the administrative borders that have existed throughout Estonia's history.

Genetic Structure and Differentiation of three Indian Goat Breeds

Kutchi, Mehsana and Sirohi breeds of goat are found in semi-arid to arid regions of North-Western India. These breeds had previously been described on the basis of phenotype and/or geographical distribution (Acharya, 1982). They are highly adapted to their environment and are an important economic resource especially to farmers of the region. The classification of these breeds based on phenotype and/or geographical distribution may not actually characterize the genetic structure of population because genetically similar individuals might be labelled differently because of distinct geography and/or phenotype. On the other hand, geographic overlap or phenotypic similarity may mask underlying genetic variation. Moreover, they share their breeding area and there may be intermixing of these breeds/populations which may cause dilution of genetic structure. Genetic variation within and between breeds is warranted to differentiate them on a genetic basis. The resulting genetic information may assist in choosing the best conservation and improvement options for these genetic resources. The need for conservation of livestock diversity and for characterization of breeds and populations, including their genetic differentiation and relationships, has been highlighted (FAO, 1995a; b). Microsatellite markers are best suited to characterize the genetic variability within and between populations because of their high variability, distribution throughout the genome, co-dominant inheritance and neutrality to selection (Boyce et al., 1996).

The present investigation was undertaken to evaluate gene flow, genetic structure and differentiation of these goat populations of North-Western India based on microsatellite loci. The individual animals were also assigned to their breed on the basis of estimated microsatellite allele frequencies.

Materials and Methods

Molecular Techniques

Blood samples (10 ml each) of 46-52 unrelated animals of each of the three breeds, namely Kutchi, Mehsana and Sirohi, were collected from their respective breeding area. The genomic DNA was isolated from these samples using a standard phenol-chloroform extraction method (Sambrook et al., 1989). A battery of 25 microsatellite markers based on the guidelines of ISAG & FAO's DADIS program was utilized to generate allelic data. Each forward primer was tagged on the 5' end with one dye out of four dyes (FAM, PET, VIC, NED) as supplied by Applied Biosystems, UK.

Polymerase Chain Reaction (PCR) was carried out using about 50-100 ng genomic DNA in a 25 [micro]l reaction volume on a PTC-200 PCR machine (M J Research). The reaction mixture consisted of 200 [micro]M each dNTP, 50 nM KCl, 10 mM Tris-HCl (pH 9.0), 0.1% Triton X-100, 2.0 mM Mg[Cl.sub.2], 0.75 unit Taq DNA polymerase and 0.4 [micro]M of each primer. The thermal cycling conditions employed were initial denaturation at 95[degrees]C for 1 min, 3 cycles of 95[degrees]C for 45 sec and 60[degrees]C for 1 min, 3 cycles of 95[degrees]C for 45 sec and 57[degrees]C for 1 min, 3 cycles of 95[degrees]C for 45 sec and 54[degrees]C for 1 min, 3 cycles of 95[degrees]C for 45 sec and 51[degrees]C for 1 min, 20 cycles of 95[degrees]C for 45 sec and 48[degrees]C for 1 min. At the end of the reaction, 5.0 [micro]l of stop dye (95% formamide, 0.25% bromophenol blue and 0.25% xylene cyanol) was added to terminate the reaction. 6 [micro]l of PCR products were run onto a 2% agarose gel, electrophoresed and visualized over UV light after ethidium bromide staining to detect the amplification.

The microsatellite genotyping was carried out using an ABI 3100 prism (Avant genetic analyzer) with LIZ 500 as internal lane standard. The data were collected and analysed using Gene Mapper Software (Version 3.0, Applied Biosystems software).

Statistical Analysis

To determine the genetic variation within and between breeds, parameters such as Nei's (1978) unbiased heterozygosity and Wright's (1978) F-statistics (F_{st}, F_{is} and F_{it}) were calculated. Heterozygosity is defined as the probability that a given individual randomly selected from a population will be heterozygous at a given locus. The statistics F_{st}, is an estimate of variation due to differences among populations, which is the reduction in heterozygosity of a sub-population due to genetic drift. The statistics F_{is} is an estimate of variation within populations that measures the reduction in heterozygosity in an individual due to non-random mating within its subpopulations. The statistics F_{it} is the overall reduction in heterozygosity in an individual relative to the total population. This includes the contribution due to nonrandom mating within sub-populations (F_{is}) and that due to population subdivision (F_{st}). The Popgene (version 3.1) program (Yeh et al., 1999) was employed to calculate the number of alleles; effective number of alleles (Kimura and Crow, 1964); observed and expected heterozygosity at each locus within and across the populations under study.

A number of statistics have been developed to summarize genetic differentiation at microsatellite loci while taking into account the incremental stepwise mutation model. We have chosen not to use these new measures as recent analyses have shown that microsatellite allele frequency drift and not mutation seems to play a more important role in genetic differentiation among closely related populations (Takezaki and Nei, 1996). The drift based classical measure of degree of genetic differentiation such as F_{st} was calculated using the FSTAT program (Version 2. 9.3.2) (Goudet, 2002). The level of significance ($p<0.05$) of F_{is} was determined by permutation test with sequential Bonferroni procedure applied over all loci (Goudet et al., 1996) to test Hardy-Weinberg equilibrium within breeds. The test for pair-wise linkage (genotypic) disequilibrium among the microsatellite loci was also done using the FSTAT program. The level of significance for genotypic equilibrium based on 18,000 permutations was calculated and the adjusted p value for 5% nominal level was 0.000056.

The effect of migration and gene flow on the genetic structure of the analysed population was estimated between each pair of populations according to an Island model under neutrality and negligible mutation (Slatkin, 1993). [N_{em}] value indicates the average number of effective migrants per generation to produce the observed [F_{st}] under the Island model. The private allele method as applied in the Genepop program (Raymond and Rousset, 1995) was used to calculate [N_{em}] estimate and private alleles.

An unweighted pair group method with arithmetic mean (UPGMA, Sneath and Sokal, 1973) was used to construct the phylogenetic tree based on Nei's (1972) standard genetic distance using the DISPAN program (Ota, 1993).

It has previously been demonstrated that the set of alleles an organism carries for a control panel of microsatellite loci can be used to identify the source population. Using the allele frequency across 25 studied loci from three breeds of goat, the individuals were assigned to their source population (breed) on the basis of simulated data based on a coalescent process by different methods using the GeneClass program (Cornuet et al., 1999). This program includes two types of methods for assigning individuals to populations. Likelihood based methods (frequency and Bayesian) assign individuals to the population in which the likelihood of their genotype is highest. Nei's (1972) genetic and shared allele (DAS) distances were used to assign individuals to the genetically closest population.

Results

The various measures of genetic variation across the three breeds of goat, namely Kutchi, Mehsana and Sirohi, investigated at each locus are present. A total of 637 alleles were observed among 146 assayed animals across the three breeds. The observed and effective number of alleles across the studied loci for the 3 breeds varied from 12 (ILST 008, ILSTS 065) to 50 (ILSTS 058), and from 4.65 (ETH 225) to 31.23 (ILSTS 058) with an overall mean of 25.48 [+ or-] 1.98 and 12.68 [+ or-] 1.32, respectively. The proportion of diagnostic alleles in Kutchi, Mehsana and Sirohi breeds was 0.053, 0.057 and 0.037, respectively. However, there were two private alleles with allele frequency of [greater than or equal to] 0.08 in Kutchi (ILSTS 019, ILSTS 30) and one each in Mehsana (ILSTS 082) and Sirohi (RM 088) breeds. The proportion of low frequency (<0.05) alleles observed at different loci in these breeds varied from 0.10 to 0.91. Shannon's information index (Lewontin, 1972), a measure of polymorphism, across these loci for the populations varied from 1.87 (ILST 008) to 3.62 (ILSTS 058) with an average of 2.69 [+ or-] 0.09. All the studied microsatellite loci were polymorphic, which indicates that the microsatellites used were suitable for genetic diversity analysis. The observed heterozygosity and unbiased expected heterozygosity over the studied loci across all three populations ranged from 0.09 (ETH 225) to 0.86 (OMHC 1) with an overall mean of 0.58+0.04 and from 0.78 (ETH 225) to 0.97 (ILSTS 058) with an average of 0.90 [+ or-] 0.02, respectively.

The gene diversity across the studied loci for the Kutchi breed varied from 0.57 (ILST 065) to 0.93 (OarFCB 304, OMHC 1, ILSTS 058) with an overall mean of 0.79 [+ or-] 0.02. The corresponding values for Mehsana and Sirohi breeds were 0.16 (ILST 008) to 0.93 (OMHC 1, ILSTS 058) with an average of 0.76 [+ or-] 0.04, and 0.50 (ILSTS 029) to 0.94 (ILSTS 058) with an average of 0.78 [+ or-] 0.02, respectively. These populations were the same based on heterozygosity. There was no differentiation between these statistics (0.76-0.79). Although varying among populations, the observed heterozygosity was lower than expected based on overall microsatellite loci for all the populations and for the pooled data across the populations. Significant deviations from Hardy-Weinberg equilibrium were detected in all the populations for most of the loci studied. Out of 25 studied loci, 22 loci showed positive deviations (observed heterozygote deficiency) in the Sirohi breed, while 13 different loci showed positive deviations in Kutchi and Mehsana breeds. Heterozygote excess (negative [F.sub.is]) was also observed for one locus in each of the populations.

Significant linkage disequilibrium (p<0.05) was detected in the overall microsatellite data for 33 locus pairs out of 300 pairs (data not shown). For individual populations, the number of tests that were significant after Bonferroni correction was: Sirohi (17), Kutchi (18), Mehsana (15). Overall means of the F-statistics obtained from jackknifing over all loci were significantly different from zero based on 't' test. The global deficit of heterozygotes across populations ([F.sub.it]) amounted to 35.9%. An overall significant deficit of subpopulation heterozygosities ([F.sub.is]) of 26.0% occurred in analysed loci. The overall genetic differentiation among the breeds

([F.sub.st]) was 13.4%. The estimates of pair-wise [F.sub.st] between each population pair revealed that Kutchi and Mehsana were more differentiated (16%) followed by Mehsana and Sirohi (13%). Kutchi and Sirohi were least differentiated (11%) among all the population pairs. The overall [N.sub.em] value of 1.95 indicated the genetic flow between the populations. [F.sub.st] values indicated a moderate level of genetic differentiation among these populations (11-16%). Takezaki and Nei (1996) showed that Nei's (1978) DA distance was generally best for inferring the correct topology for the infinite mutation (IAM) model. Thus, DA was used to derive a dendrogram of genetic relationship among the populations, constructed as a neighbor joining tree (Saitou and Nei, 1987). However, the tree nodes had very low bootstrap values (9-16%) and consequently the tree collapsed.

The measures of standard genetic distance between pairs of breeds indicated that the lowest genetic distance was between Kutchi and Sirohi (0.73) and that the largest genetic distance was between Mehsana and Kutchi (1.0) followed by Sirohi and Mehsana (0.75). With the aid of prior population information, 25 microsatellite markers could assign the individuals to their respective breed 65 to 83% of the time using likelihood based methods of breed assignment. The test, in essence, resolved whether an individual of a particular breed possessed a genotype that is typical of its own breed or whether it better reflected the genetic characteristic of another breed. On the other hand, genetic distance-based methods could assign the individuals to their respective breed 44 to 91% of the time.

All measures of genetic diversity revealed that there was substantial genetic variation within and across the three breeds studied. The substantial genetic variation has also been observed in many other breeds of goat. Although varying among populations, observed mean heterozygosity was lower than the expected mean heterozygosity for all the populations. The majority of the loci (>50%) in each population showed heterozygote deficiency, which may be due to presence of low frequency alleles segregating at many of the loci and the significant genotypic linkage disequilibrium observed in these breeds.

The moderate level of genetic differentiation (Fst) among the three breeds (13.4%) under study implied that 86.6% of the total genetic variation corresponded to differences among individuals within breed and 13.4% of the total genetic variation corresponded to the breed. This level of differentiation among the populations is within the range of Fst values reported in the literature. A value of 0.146

was found in indigenous goats of sub-Saharan Africa (Chenyambuga et al., 2004); 0.105 in 12 Chinese indigenous goat populations (Li et al., 2002); 0.17 in Swiss goat breeds (Saitbekova et al., 1999); 0.073 in Italian goat populations (Iamartino et al., 2005); 0.202 in goats from Korea and China (Kim et al., 2002); and 0.227 in Asian goats (Barker et al., 2001). The overall [F.sub.is] value (0.26) was higher and significantly different from zero indicating departure from random mating and suggested that some of the studied loci were homozygous in the populations. Some of these homozygous loci could be undergoing natural selection or linked to other loci affecting morphological, productive or adaptive traits undergoing natural selection. The measures of genetic differentiation revealed that the different populations remained genetically distinct, despite the fact that there is no breeding policy to create breeds or maintain the breed purity.

With prior information, the assignment of individuals to the rightful breed based on 25 microsatellite markers revealed that different methods used could not effectively discriminate individuals with a moderate level of differentiation among the breeds determined by their history of development, breeding and management practices to their rightful populations. Our results are in accordance with the results of Awemu and Erhardt (2005) for zebu cattle breeds.

In conclusion, all the measures of genetic variation revealed substantial genetic variation in each of the studied populations and thereby there is good scope for their further improvement. The measures of F-statistics differentiated the breeds moderately on a genetic basis and revealed a significant level of heterozygote deficit in the populations at some of the studied loci. The measurement of gene flow showed immigration of new genes from other populations and thereby intermixing of the populations; hence, this may pose a danger to the purity of the breed in the long run. Therefore, suitable breeding strategies are required for maintaining breed purity and improving these breeds

Study on Genetic Evaluation for linear Type Traits in Holstein Cows

During the past ten years, the number of dairy cows in South Korea declined by 90,000 from 550,000 to 460,000 while the number of dairy operations declined 33%, from about 24,000 to roughly 8,000. However, because milk yields per head have increased 25% (from 6,000 kg to 7,500 kg), the total lifetime milk yield was not affected, and this indicates increased performance and thus increased physiological stress on the cows. This physiological stress can have negative effects on other aspects of milk production. For example, the reproductive period increased 40 days from 400 to 440 days during the 10-year period reported previously, and, in 2005, of the 158,000 cattle tested, the animals usually considering the most productive (parity 3 to 5) comprised only 34% of the total cow herd with an average parity of 2.4 (Korean Agricultural Cooperative Federation, 2006), perhaps indicating decreased longevity of the animals. In addition, even though total milk yield is increasing, management costs are also increasing greatly, decreasing the efficiency and profitability of the industry. To solve these kinds of problems in other countries, the dairy industry has been making efforts for better economic proficiency of the population by selecting individuals with genetically excellent reproductive performance and body type, better udder development and increased production longevity.

Linear type traits can be used to maintain a more profitable herd of Holsteins through the selection of better bulls. Understanding the trend of linear type traits can assist with the selection of a better group of mating sires. The Holstein Association USA has been testing linear type traits of cows, including fifteen basic linear type traits and total scores such as general stature composite (GSC), dairy capacity composite (DCC), body size composite (BSC), foot and leg composite (FLC), and udder composite (UDC) as a relatively comparative test among individuals. In South Korea, the Korea Animal Improvement Association (KAIA) has been responsible for testing using the USA body scoring system since 1980. Collected data has been used for selection of seedstock. Definitions and collection methods of scoring linear type traits have been reported by Song et al. (2002). The actual number of dairy cows tested by KAIA was about 32,000 in 2005, which is equivalent to 11.9% of total dairy cows. Dairy farmers in South Korea have been trying to select high-performing cows over multiple parities through comparative tests within contemporary groups; however, few studies involving genetic analyses of these traits have been conducted, likely due to the difficulties in estimating genetic parameters for animals with data collected with the comparative scoring method. The method used ranks the traits with one of six scores: poor, fair, good, good plus, very good, and excellent.

The development of alternative estimates of genetic performance would provide another selection tool for Korean milk producers and would also allow for comparison of animals by genetic country of origin. Therefore, the objectives of this study were to i) determine alternative methods to evaluate genetic performance for linear type traits of individual Holstein dairy cows, especially focusing on comparative traits, and to estimate genetic variances for these traits using actual data, and ii) compare genetic performance and improvement of progeny by birth country of the cows.

Material and Methods

Data

Data used in this study was collected from 118,290 Holstein dairy cows between the years of 2000 and 2004 by KAIA. Based on KAIA testing program rules, comparative traits were investigated on milking cows within herd without taking into account age and stage of lactation. Only data with more than five tested cows by herd appraisal date and of sires having more than ten daughters were included to increase the reliability of the data analyses. Outliers were removed and the records of individuals having data from one or both parents were included in the analyses. The data used in the analysis is a total of 30,204 records of the selected traits, which was collected from 26,701 individuals having pedigree information.

Five comparative traits were investigated in this study, general stature composite (GSC), dairy capacity composite (DCC), body size composite (BSC), foot and leg composite (FLC), and udder composite (UDC). These traits were scored from 1 to 6 by trained technicians with 1 = poor, 2 = fair, 3 = good, 4 = good plus, 5 = very good and 6 = excellent. Final scores (FS) also were determined with a range of 50 to 100 considered the linear type traits.

Statistical Model

An analytical model was set based on a previous study designed to focus on estimating genetic parameters for these traits by threshold models using Gibbs sampling algorithms (Lee, 2006). Herd appraisal date, year of age, lactation stage (grouped by month), and time lagged for milking (in hours) were assumed as fixed effects in the model. Animal additive genetic effects considering pedigree relationship and residual errors were assumed with random effects. Year of age at appraisal date was classified from one to nine years of age, with the value of nine years of age assigned for animals that were greater than or equal to nine years of age. Lactation stage was classified by month from one to 12 months after parturition. Times of appraisal were classified into one of 12 h, using hourly intervals between milking and appraisal time. Statistical model was as follows:

Due to non-observed latent variables for type classification traits, genetic evaluation for these traits as well as final scores with assuming linear trait was adopted on a linear-threshold model using Empirical Bayes theorem providing by Quaas (2001) with multiple trait animal models. In this study, genetic parameters were used to estimate breeding values for these traits. This model was a linear-threshold model by likelihood approach, and was a conversion from BLUP90IOD (Misztal, 2008) using formulas given by Quaas (2000).

Results and Discussion

Genetic parameters estimated in this study are provide. Tsuruta et al. (2004) reported heritabilities of 0.40 to 0.49 in BSC, and 0.26 to 0.38 in UDC for analysed linear type data from the Holstein Association USA with multiple-trait animal models using random regression. The heritability value for the present study for BSC (0.473) falls within the range reported by Tsuruta et al. (2004) but the UDC heritability value for the present study (0.430) is slightly higher than the value reported by Tsuruta et al. (2004). The genetic correlation between BSC and UDC reported by Tsuruta et al. (2004) was weak, estimated at-0.15 to 0.04, in contrast to the correlation of 0.368 noted in the present study for BSC and UDC. The heritability of BSC has been reported as 0.19 for first lactation cows and 0.22 for all cows, which was analysed with ASREML using sire models with the data from the Holstein Association USA (Dechow et al., 2003) while the heritability for BSC in the present study was 0.473. However, in these analyses, the effect of age at calving nested within lactation, 5th order polynomials of DIM, fixed herd-classification visit effects and random sire and error, etc. were included.

Genetic resources of Holstein in South Korea have been imported for several decades. Imported countries of genetic resources were identified in this study, and the breeding values by imported countries were compared, potentially providing a valuable reference for farmers who might be considering the importation of semen from abroad. The breeding values with standard errors as estimated by country of importation are represente. The US had the largest set of data, followed by Canada, Japan and Australia. Among all countries, cows imported from Canada had greater breeding values for GSC, DCC,

UDC, and FS, those from Australia had greater breeding values for BSC, and those from Japan had greater breeding values for FLC. The greater UDC breeding values from genetic sources imported from Canada is similar to that reported by Park et al. (2006).

Powell et al. (2003) reported EBV of conformation traits in several countries, with mean sire EBV for stature estimated at 0.87,-0.03 and-0.17 for Canada, Australia, and the USA, respectively. The authors also estimated mean sire EBV of fore udder and udder support as-0.45 and 0.09 in Australia,-0.06 and 0.40 in Canada, and-0.09 and-0.12 in the USA, respectively.

In the present study, the greatest breeding values for GSC were estimated for Canada with the breeding values for American lines increasing for 10 years starting in 1989 but tending to the decrease after that until 2004. For DCC, the breeding values for American and Canadian lines showed similar patterns until 1999, after which the breeding values for the American lines declined sharply. For BSC, data from Korea, Canada and the USA followed similar trends, until the breeding values of the Amercian lines started decreasing in 1999. In the US, selection indexes were based primarily on milk yield traits until methods for evaluating other traits began to emerge, likely explaining this phenomenon. In 1994, selection indexes for somatic cell score and productive life were added, and in 2000, a composite type index for body size selection was added (Shook, 2006). The selection of a smaller, more feed efficient animal that would also maintain high a productive life would be beneficial to the dairy industry. Therefore, it is possible that American producers began to use productive life and body size indices for genetic selection, resulting in a doubling of downward selection pressure on animal size (Shook, 2006). Thus, the decrease noted in the present study for GSC, DCC and BSC breeding values (which could all be related in part to animal size) for American lines of Holsteins in South Korea from 1999 to 2004 would be expected.

Breeding values of FLC were estimated to be highest for lines from Canada in 1989, however, there was a similar pattern for Canada, Korea, and the USA after 1989 and up to 1999, with American lines having greater breeding values after 1999 compared to Canada and Korea. The higher breeding values for the USA lines after 1999 are probably linked to the introduction of the foot and leg composite selection index to allow for greater selection of this trait in breeding animals starting in 2000 (Shook, 2006). However, an udder composite index for genetic selection was also added in 2000 for USA breeders (Shook, 2006), and since udder trait selection is important in dairy cattle, it is surprising that the breeding value decreased from 1999 to 2004 for the American lines for UDC. However, genetic selection is driven by the bulls provided for producers to choose from as well as their choices of traits for which they will base their selections, including production indexes. If selection of Holstein sires from America by South Korean producers emphasized traits negatively related to UDC in the years up to 1999, that might partially explain the decrease in UDC from 1999 to 2004 in the present study.

From 1984 to 1994, UDC breeding values of the three countries analysed followed a similar pattern, however, those of Canada increased rapidly after 1994 and though South Korea and USA values were similar until 1999, breeding values for American lines decreased after 1999, as previously noted. Breeding values of FS were greater for Canadian lines starting in 1984, and those for USA lines decreased after 1999. Overall, these data provide valuable information for scientists and producers involved in genetic selection of dairy cattle in South Korea and, in future years, more analysis should be conducted to continue to follow trends that could impact the dairy industry.

New Ultrasound and Artificial insemination Techniques Improve Sheep Breeding

Bred for meat as well as wool, sheep have been part of the American landscape since colonial times. Now, advances in genetic research are changing breeding procedures, leading to faster, cheaper, and more accurate techniques.

Using Ultrasound for Selective Breeding

New research is making it easier to predict the future—at least as far as lamb products are concerned. At the U.S. Sheep Experiment Station in Dubois, Idaho, research leader Greg Lewis and his colleagues have shown that ultrasound can be used to accurately predict characteristics that indicate carcass yield and value in live sheep.

This could significantly improve the speed and accuracy of selective-breeding methods. Since carcass data is difficult for producers to obtain, many of them currently rely on visual appraisals to predict carcass traits before choosing which sheep to breed. Ultrasound provides a faster, more accurate alternative. To assess the reliability of the technology, scientists captured ultrasound images of 172 lambs before slaughter. At Ohio State University Ohio State University, main campus at Columbus; land-grant and state supported; coeducational; chartered 1870, opened 1873 as Ohio Agricultural and Mechanical College, renamed 1878. There are also campuses at Lima, Mansfield, Marion, and Newark., assistant professor Henry Zerby coordinated the collection of carcass-trait data for the lambs. Lewis collaborated with ARS scientists and Dave Notter, a geneticist ge net i cistA specialist in genetics.geneticist a specialist in genetics.geneticist at Virginia Tech, to analyse the data.

Results showed that a trained technician can capture an ultrasound image in about 30 seconds with reasonable accuracy. Scientists can use the images to estimate traits that influence the carcass value of market lambs—such as loin loin (loin) the part of the back between the thorax and pelvis. loinThe part of the body on either side of the spinal column between the ribs and the pelvis. muscle area, loin muscle depth, and back-fat thickness.

Ultrasound is initially more expensive than visual appraisals, but the technique's superior accuracy may translate into better economic returns through improved evaluation and selection of breeding stock.

"Ultrasound is a great way for breeding-stock producers to get the data they need to make selection decisions," Lewis says.

Reliable predictions are timesavers for breeders because they enable them to make educated decisions about an animal's offspring without waiting for the offspring to mature. While the technology is used routinely for cattle and swine, this study demonstrates that it can also be applied to sheep.

Improving Surgical

Artificial insemination (AI) is an important tool for modern livestock breeding—particularly for cattle and swine. But for most sheep producers, AI isn't a viable option. The ovine ovinepertaining to, characteristic of, or derived from sheep ovine atopic dermatitis symmetrical erythema, alopecia, lichenification, excoriation on woolless areas; sporadic cases, recur each summer. cervix is relatively difficult to traverse with traditional AI tools, so techniques that work easily with cattle have been less successful with sheep. Surgical AI has a higher success rate, but is prohibitively complicated and expensive.

But help is on the way, thanks to research conducted by ARS animal scientists at the National Centre for Genetic Resources Preservation (NCGRP NCGRP National Centre for Genetic Resources

Preservation (United States Department of Agriculture)) in Fort Collins, Colorado The City of Fort Collins, a home rule municipality situated on the Cache la Poudre River along the Colorado Front Range, is the county seat and most populous city in Larimer County, Colorado.. Researchers there are collaborating with scientists at the University of Wyoming UW is a national research university prominent in the fields of environment and natural resource research, specializing in agriculture, energy, geology, and water resource related fields. (UW) to improve semen handling and artificial insemination on sheep farms. The work is being led by NCGRP animal physiologist Phillip Purdy in collaboration with animal geneticist Harvey Blackburn and UW colleagues Robert Stobart and Brent Larson.

"Effective AI has two components: the ability to collect and transport semen—either fresh or frozen—to producers around the country and the ability to economically inseminate in seminate. To introduce or inject semen into the reproductive tract of a female.

The scientists first focused on semen collection, storage, and transport. They found that sheep semen can be collected and shipped—in cooled liquid form—overnight before freezing without harming its in vitro quality. They also found that semen could be cryopreserved—or frozen for later use—after shipping without reducing its success in surgical AI.

Further studies compared semen that had been cryopreserved immediately after collection to semen cryopreserved after 48 hours—about the time required to ship samples across the country. The scientists observed no difference in quality or fertilization ability nor in the average number of lambs born to the ewes surgically inseminated with those samples.

"These results show that there are many options available to producers to help them select for desirable traits," Purdy says. Producers could, for example, use samples from around the world, thereby expanding and improving their breeding options, increasing the quality of their flocks, and providing a better return on their investment.

This work also has benefits for conservators of genetic resources, like the NCGRP, that gather and store genetic materials. "An efficient insemination insemination the deposit of seminal fluid within the vagina or cervix artificial insemination (AI) that done by artificial means method is important for our conservation efforts," says Blackburn, who heads NCGRP's National Animal Germplasm Program.

"If we can do it more efficiently, we can reduce the number of semen samples we collect for preserving sheep genetic diversity."

Better Nonsurgical AI

The scientists have developed a rapid, economical alternative to existing surgical AI methods. The technique is an adaptation of the method used in swine. The researchers used a spiral insemination catheter to traverse the ewe's cervix and deposit thawed semen directly into the uterus. The method is easy to learn and easy to perform. Each sheep takes about 2 minutes to inseminate at a cost of $1.29, making it significantly faster and less expensive than laparoscopic surgical insemination. And it's easy enough for producers to do independently.

Though quick and economical, the technique is less reliable than existing AI methods. Early tests have had success rates of about 55 percent when using fresh semen and about 10 percent when using frozen semen.

"This difference was not observed in the previous experiments, because surgical inseminations deposit the AI dose at the site of fertilization, so minimal stress is placed on the thawed sperm," Purdy explains. "Sperm are generally weakened by the freezing and thawing processes, and this impaired function in thawed samples becomes more noticeable when using the nonsurgical method because the sperm have to travel farther to the site of fertilization."

Meeting Nutrient Requirements

A young, growing heifer at approximately 600 lb will require about 10.5-11.0 per cent crude protein and 0.45 Mcal/lb of energy for gain. These numbers may seem high unless one accounts for increased maintenance due to cold temperatures. In the upper Midwest, many heifers, due to snow cover, are fed in the dry lot using harvested forages. This typically includes some type of dry hay, a grain source, and perhaps an ensiled feedstuff. Often, the dry hay is at least 12 per cent crude protein and even moderate provision of a grain source will meet energy demands.

In contrast, if an ensiled feed is used, energy needs will be easily met by the silage. This will produce a gain of 1.5 lb/day, which is sufficient to produce a heifer of adequate size for breeding by spring. This is especially true when one considers compensatory gain. Compensatory gain is the type of gain experienced by a previously restricted animal when exposed to a more nutrient dense environment, such as spring/early summer pasture. Perhaps of greater concern in areas of the country where weather necessitates extended dry lot feeding is creating heifers with excessive body condition. Older data in dairy type heifers indicate that heifers weighing 30-35 per cent more than lighter counterparts at calving produced 700 lb less milk during their first lactation (Swanson, 1960). This was likely due in part to inhibited mammary development in fat heifers.

This observation is supported by Ferrell et al. (1976). Ferrell offered pregnant heifers a high or low level of nutrient intake. Heifers were slaughtered and those on the high level of intake had greater udder weight and a larger proportion of that weight was fat. In addition, heifers developed to weigh 700 versus 630 lb at breeding required more services per conception, which may suggest an increase in embryonic mortality for heavier heifers (Short and Bellows, 1971). Arnett et al. (1971) studied a more substantial difference in prebreeding weight compared to Short and Bellows (1971).

Sets of twin heifers were developed at a rate of 0.75 lb/day or at a rate designed to induce "a high degree of body fatness" (Arnett et al., 1971). The overdeveloped twins were 30 per cent heavier than more appropriately developed twins at first mating and essentially maintained this difference through their third calving season.

Developing heifers to become obese increased the number of service required for a conception by 7 per cent compared to leaner heifers. This difference persisted through the third breeding season. More striking was the difference in calving difficulty, where obese heifers required 86 per cent more assistance for calving than lean heifers and this difference also persisted through the third calving season. Across three lactation periods, heifers developed to become obese produced less milk than lean counterparts did and as a result, their calves were lighter at weaning. The difference in milk production is further supported by data from Ferrell (1982).

Heifers developed at 1.75 lb/day produced 1.20 lb/day less milk than heifers developed at 1.30 lb/day. Furthermore, the heavier developed heifers weaned 14 kg lighter calves. Patterson et al. (1992) cited numerous other studies that provide additional support for the concept that developing heifers to become obese negatively affects reproduction and calf production. However, more importantly, developing heifers to excessive weights is expensive and requires additional dollars per each heifer developed. Postweaning management of heifers to achieve traditional target weights, particularly by feeding high-energy diets, is not supported by current research. Heifers developed on forage, however, generally require additional protein supplementation to achieve even modest gains. One reason reproductive performance has not been drastically impaired by feeding to lower target weights may relate to genetic changes in age of puberty. While the systems presented herein are not directly applicable to the upper Midwest, the principles of low input development are if used correctly. Regardless of the system, minimising expensive feedstuffs will reduce development cost, which is a major determinant in lifetime cow profitability.

Strategic Crossbreeding

For cow-calf producers, survival and operational profitability are reliant upon efficiently producing uniform calves for target markets in an economical fashion. Doing so requires a clear management plan, set goals for the cow herd, proper bull selection and a concise marketing strategy. Collectively, those things reduce risk and generate greater returns to the bottom line." Breeding cattle isn't rocket science, although it does require common sense and a certain degree of business savvy," said Mike Horvath, director of commercial marketing for the North American Limousin Foundation (NALF). "Simply put, there are three basic markets available: mainstream, lean and premium-Choice. The majority of commercial producers shoot for the mainstream commodity market, with a handful of managers gearing their enterprises towards the lean or premium-Choice markets."

For the latter two options, he explained, producers must be comfortable accepting greater risk. For example, if a calf destined for a natural program gets sick and requires an antibiotic, its destiny changes.

Despite the target market, true success starts in the cow herd." Regardless of breed composition, cows must be sound-structured, big-ribbed, loose-flanked, easy-fleshing and moderate," Mr Horvath explained. Conservatively sized females usually will be more profitable and efficient because they often wean more total pounds of calf per cow exposed, and their calves have more market flexibility as they either can be backgrounded or go straight into the feedyard. Additionally, females should be genetically balanced – using expected progeny differences (EPDs) – if they are to be productive in the herd.

Furthermore, producers realise additional value through maintaining a crossbred cow base. Production and economic advantages of commercial crossbred cows, adapted to their environment, will trump those of straightbred cows, with the following advantages:

- 20 per cent more pounds of calf weaned per cow exposed and;
- an additional 1.3 to 2.0 years of cow longevity.

"That equates to a 30 per cent increase in lifetime cow productivity," Mr Horvath stated.

In terms of dollars and cents, an average commercial F· cow will return about $70 more per year than a straightbred. If the chosen crossbreeding system yields between half and two-thirds of maximum hybrid vigour (heterosis), the additional $50 per cow per year yields at least $400 more in lifetime earning over a straightbred.

With a strong cow base, managers have more versatility in bull selection and, ultimately, greater access to a larger array of market opportunities. Producers should apply many of the same criteria used in female selection when seeking out bulls.

"Potential sires should be big-footed, good-structured, high-capacity, muscular and have above-average testicular development," Mr Horvath advised. "Genetic composition and associated EPDs should be above breed average and progressive for the direction the enterprise is headed."

For those targeting the mainstream market, NALF recommends producing halfblood calves by turning out purebred Limousin bulls with British-based cows then through the use of F· Lim Flex® bulls on the F· Lim Flex females produced.

"The advantage of 'Lim Flex on Lim Flex' is that it allows producers to realize the benefits of hybrid vigour, alleviates many of the requirements of traditional crossbreeding schemes and allows for easy retention of commercial heifers," Mr Horvath said.

"Breeding hybrids to hybrids also affords managers a heightened degree of consistency and predictability in producing uniform calves, resulting in increased merchandising ability."

NALF encourages those aiming for lean markets to incorporate higher percentage Limousin genetics so the resulting calf crops are at least 75 per cent Limousin influence. That is simple to do by pairing purebred Limousin bulls with halfblood or three-quarterblood Lim Flex cows or by breeding straightbred Limousin bulls to Limousin cows.

3

Animal Breeding and Genetics

The development of selection index theory 70–80 years ago formally embedded economic principles in the practice of animal breeding, and many of our leading practitioners have made important contributions in this area. Animal breeders have relatively infrequently drawn on economic theory and instead have developed their own simple and practical tool kit to address the issues of prioritizing traits for genetic selection decisions and when undertaking cost benefit analysis. We know from our quantitative work that models do not have to be true in order to be useful, and the tool kit we have has certainly proved useful. It is also much more accessible than many of the highly complex theories that exist in economics.

In this editorial, I argue that principles of economics can help us to better understand successes and failures in animal breeding. Notable successes include the massive and sustained improvements in the productivity of intensively farmed livestock species, particularly pigs and poultry systems but also dairy production systems. Failures include attempts to move exotic genotypes into unsuitable environments, and the relatively low return to substantial investment in ongoing genetic improvement in extensive and low-input production systems in both the developed and developing worlds.

When compared with most farm management alternatives, genetic improvement is an extremely slow and quite clumsy technology option. For many of our successes, these deficiencies have been offset by the permanent and cumulative nature of genetic improvement, and the ability to remould farm management systems and practices to suit the new genotypes that have evolved.

The main farmed species and farming systems are highly varied in terms of how genetic improvement can be achieved and distributed, and undoubtedly, these differences have contributed to the successes and failures. However, economic principles of market failure, vertical integration, value capture, intellectual property (IP) rights, efficiency because of competition and disruptive technologies interact with these differences and so are also key contributors to success. A relatively simple understanding of these factors can be useful in identifying potential remedies for our not infrequent failures. In the absence of this understanding, we as animal breeders may be tempted to continue serving up the same old approaches using ever more complex statistical methods and highly expensive genomic tests that fail to free industries from the shackles of low adoption and ineffective distribution of genetic improvement.

Attempts to address market failure in animal breeding have a long history, most notably in the establishment of novel and distinctive physical characteristics (horns, ears, coat colour, body shape, etc.) which effectively brand breeds and sometimes lines within breeds such that they can achieve a market premium. Differentiation relative to unimproved commercial or cross-bred stocks is critical, but differentiation relative to lower-performing competing breeds and lines within breeds is also very important.

Unfortunately, there is also a long history of selection focus drifting almost fully onto distinguishing characteristics, and a consequent loss of focus on the production characteristics required for ongoing improvement. Within breeds, failure of commercial farmers to be able to physically observe superior genetic performance, either at all or within a reasonable time frame, has been and still is a major impediment to improvement in many extensive livestock farming systems. It seems no coincidence that the least successful livestock species for within-breed improvement are commonly those with the most extensive farming systems.

Vertical integration, whereby a single commercial entity maintains ownership of animals from breeding through to processing, and even marketing and distribution in the livestock value chain, often overcomes many problems of market failure. Improved animal identification, measurement technologies and information systems are also helping to create more transparent markets. This is important, because vertical integration usually requires large scale to justify a breeding programme. Increasing scale also increases risk. Vertical integration is often but not always a key contributor to the success of intensive livestock breeding programmes. Unfortunately, extensive livestock industries are already subject to variation in the production environment. Globalization of markets appears to be exacerbating fluctuations in input and output prices. It seems therefore that vertical integration is unlikely to accelerate rapidly in extensive livestock industries, because of the concentration and magnification of risks.

Value capture is a critical contributor to the success of many plant breeding programmes. For example, the inability of farmers to multiply their own seed from hybrid corn creates a captive market for seed, which provides revenue for ongoing investment in genetic improvement. Value capture is also a key and often highly controversial factor in animal breeding programmes. The massive and rapid international dissemination of elite Holstein dairy cattle semen leaves AI companies on a desperate treadmill striving to maintain a point of difference, while their competitors' opportunities to identify elite sons of the top sires globally have historically been limited only by the relative sizes and efficiency of their respective progeny testing schemes. But still, a historically strong emphasis towards the need for high reliability, itself a conservative measure of risk compared with accuracy, has helped dairy AI companies to stave off potential competition from semen from high-merit unproven young bulls.

The trend in animal breeding programmes over many decades has been towards privatization of breeding programmes, and a switch in publicly funded research effort away from practical breeding and towards a search for new step change and/or disruptive technologies. IP rights pertaining to germplasm are weakly controlled in animal breeding. This motivates prevention of leakage of elite germplasm to competitors, something that has been achieved to a reasonably high extent in pigs, poultry and for some species in aquaculture. This protection seems justified in the context of the developed world, relying on private investment to drive high levels of genetic progress.

In contrast, for developing countries, much of the genetic progress that arises comes about through lateral shifts and exploitation of high-performing germplasm. Indeed, for much of global dairy production, and in extensive livestock industries, genetic progress also comes about through the same means. These lateral shifts occur at the levels of sire sharing within group breeding schemes up to wide-scale international trade in elite germplasm across many species and farming industries. Thus, competition has led to more efficient breeding programmes in some industries, but sharing of germplasm is a substantial contributor to a large amount of realized genetic gain in others. Molecular biologists have long-threatened animal breeders with new approaches that would act as disruptive technologies for the process of achieving animal genetic improvement. After many years of promises, genomic selection appears to be on the verge of offering a major disruption to the global business of dairy cattle breeding. Barriers to market entry by new progeny testing initiatives have revolved around the cost of setting up a scheme sufficient to meet minimum reliability criteria and the time required to achieve results in industry, in an already highly competitive business. These barriers can potentially be broken down by public investment and availability of accurate predictions of the genetic merit of young bulls. It will be interesting to see how the future unfolds. Genomics may yet help reverse the trends of disappointing rates of genetic improvement in extensive species and in developing countries.

Big decisions and consequent changes to approaches are likely for animal breeding over the next decade as food demand grows and new technologies continue to unfold and be exploited. Economics, and an understanding of the past, will help us to shape the future to take full advantage of the huge potential of livestock genetic improvement.

Genetic Variability of Traits

The objectives of this study were to estimate direct heritability of 30 conformation and performance traits evaluated during 100-day stationary performance test and to predict the genetic effects in Polish Warmblood stallions. Inbreeding coefficients were estimated as well. Moreover, phenotypic and genetic trends were derived. The data were extracted from the database of the Polish Horse Breeding Association. The analysis included 494 warmblood stallions, which performed during 100-day test in the years 2002-2008 at two Polish Training Centres. Pedigree data comprised 8512 individuals. Restricted maximum likelihood was employed to estimate parameters under an additive genetic model (including fixed effects: age of stallion, breed, year and place of performance test). Nonzero inbreeding coefficients were estimated for 88 stallions with performance records as well as for 458 unrecorded ancestors. Average inbreeding level for the stallions with records was 0.29%, whereas for all inbred individuals – 1.69%. The average completeness of the pedigrees, expressed as Cassell coefficient, for the recorded stallions was 42.47%. The heritability estimates varied from 0.14 (character) to 0.87 (total index). Relatively high heritabilities were found for jumping-ability traits. In general, the genetic trends for studied traits were negligibly positive.

Calving records from the Animal Breeding Centre of Iran collected from January 1990 to December 2007 and comprising 207106 first calving events of Holsteins from 2506 herds were analysed using linear sire models to estimate genetic trends for age at first calving (AFC) and milk traits. Genetic trends were obtained by regressing yearly mean estimates of breeding values on year of birth. In general, there were decreasing genetic trends for AFC, fat percentage and protein percentage over the years but there were increasing genetic trends for milk yield, mature-equivalent milk yield, fat yield, mature-equivalent fat yield, protein yield and mature-equivalent protein yield over the years.

On the other hand, there was a decreasing phenotypic trend for AFC but estimates of phenotypic trends were positive for milk yield and compositions over the years. It seems that the decline in calving age in this study over time resulted primarily from increased turnover rate and increased culling of heifers that failed to get pregnant. On the other hand, increasing trend for yield traits over time in this study indicated that Iranian dairy producers were successful in choosing progressively better semen and sires from imported and local sources over the years.

The main objective was to locate QTL and estimate the proportion of total genetic variance attributable to quantitative trait loci (QTL) for production index traits and the udder health index identified on six Bos taurus autosomes in the Danish Holstein dairy cattle population. Data were obtained from a granddaughter design of 20 sire families with a total of 1869 progeny tested sons. The number of sons per grandsire ranged from 20 to 284, with an average family size of 93.5. Indexes of the estimated breeding values were obtained for the milk production traits and for the udder health index from the Danish Agricultural Advisory Service database. A random-QTL model was applied to incorporate marker information into parameter estimation for each single QTL. The procedure allowed us to detect new QTL on BTA3, BTA16 and BTA28 and to estimate the proportion of total genetic variance attributed to different QTL on a total of six Bos taurus autosomes for the udder health index and yield index traits in the Danish Holstein population. Variance estimates vary between 2 to 58% of the total variance for different QTL and seem to explain a substantial part of the variance at certain positions of the cattle genome. The results are discussed against the background of the failure of marker-assisted selection (MAS) and the recent availability of large panels of single nucleotide polymorphisms (SNPs) that have improved the search for mutations underlying variation in complex traits resulting in modern genomic selection.

This study was carried out to investigate a new quarter-individual milking system called MultiLactor (Siliconform GmbH, Türkheim, Germany). The MultiLactor enables milking on quarter level basis with low vacuum (37 kPa), sequential pulsation and periodic air inlet. Within the same dairy farm, the influence of this quarter-individual milking system (MULTI) on milkability traits was compared with a conventional milking system (CON). CON was equipped with a conventional milking cluster and used alternating pulsation. Vacuum level was adjusted to 40 kPa. For the study, 84 Holstein Friesian cows were randomly selected and uniformly divided into two herds.

During the 30-week survey, the milk flow curves were recorded every other week by using a LactoCorder (WMB, Balgach, Switzerland). Significant differences (P<0.05) between both milking systems were found for all milk flow traits, except for milk yield and decline phase. Concerning the incline (tAN) and plateau (tPL) phase, large differences existed between MULTI and CON. The estimated value of tAN calculated for MULTI (29.4 s) took only half of the time when calculated for CON (56.4 s). The estimated value of tPL at CON was reduced by 1.43 min (35%) compared to MULTI. Milking process at MULTI (8.49 min) took longer time than for CON (7.43 min). From the study, it was concluded that the effect of shorter tAN in the quarter-individual milked cows may be related to additional prestimulation by an actuator. In contrast, the longer milking time in MULTI is possibly caused by lower vacuum level and periodic air inlet.

The objective of this study was to examine semen quality parameters for the Pøeštice black-pied breed over a test period of 8 years while considering the potential effects of collection month and boar age. Ninety-nine ejaculates were collected using the gloved-hand technique from healthy and fertile mature boars from selected farms. Ejaculate volumes were relatively low because the boars were accustomed to natural mating. Sperm motility, sperm concentration, percentage of morphologically abnormal spermatozoa (MAS), and sperm motility after 24 h of storage in Androhep extender (dilution rate 1:1.5) were assessed. Significant differences were found in sperm concentration and MAS rate in relation to collection months and boar age in the monitored years (P<0.05). A tendency for MAS to increase with monitored years was observed. Significant differences in sperm motility and motility after 24 h of storage were only observed in relation to collection months (P<0.05). Results of this study detected effects due to collection month and boar age on boar semen quality during the monitored years.

The objective of the present study was to determine the polymorphism in the prolactine receptor (PRLR) gene in Chios, White Karaman and Awassi, which are native sheep breeds in Turkey. By means of PRLR gene sequence homology between sheep and humans, two primer pairs were designed for polymerase chain reaction (PCR) amplification within intron 1 and exon 10 of the PRLR gene in sheep. A total of 160 amplicons (99 for intron 1 and 61 for exon 10) were subjected to DNA sequence analysis. For intron 1, 6 different haplotypes were determined. For exon 10, 7 different haplotypes were obtained. Some variations determined for exon 10 (g.14A>T p.Q14L; g.160G>A p.D160N; g.166G>A p.E166K; g.167A>T p.E167V; g.176A>T p.H176L; g.206G>A p.S206N; g.208G>A p.G208R) led to changes in the amino acids, but no amino acid changes were determined in g.2A>T, g.81A>G, g.138A>G, g.186C>T, g.207T>C. It was noted in particular that White Karaman and Awassi were similar to each other in both PRLR exon 10 and intron 1 haplotypes, whereas the Chios breed had a different variation.

The aim of the presented study was an estimation of the zinc bioavailability derived from amino acid complexes with methionine, lysine and glycine in growing lambs. 48 lambs, Polish Merino × Romanowski × Charolaise crossbreed, at age about 10 weeks and average body weight of 20 kg were randomly divided into 4 experimental groups. Animals were kept collectively and fed with concentrate and hay in amount: 0.7-1.2 and 0.3-0.5 kg/day/head, respectively. After rearing period 6 rams from each group with average body weight of about 30 kg were divided into digestibility-balance experiments. At the end of the experiment the blood samples were taken, then from each group 8 lambs were chosen, killed and during dissection tissue samples were taken (liver, pancreas, kidney and rib bone).

In tissue samples as well as in fodders and excrements the content of zinc was determined. Obtained data in digestibility experiments allowed on calculation of apparent absorption and retention of zinc and the level of Zn determined in tissues' samples were used to the estimation of bioavailability of zinc from different ones applied in experiment sources. Apparent absorption and retention of zinc were higher (Pd"0.01) in lambs receiving in mixtures amino acid complexes of zinc than in animals from the control group. The higher content of zinc in soft and hard tissues of lambs which received organic forms of zinc might indicate better assimilation and bioavailability of zinc from these forms than from oxide. Among tested organic forms of zinc, the zinc-lysine complex was characterized as having the most advantageous property.

The aim of this study was to assess the seasonal trend of the serum protein content of sheep and goats. The tested animals were six female goats (Maltese breed) and six female sheep (Valle del Belice breed). All animals were clinically healthy and not pregnant or lactating before or during the study. On blood samples, collected through an external jugular venipuncture every 30 days for 12 months, electrophoresis was performed using a semiautomated AGE system and then electrophoretic curves with the relative protein concentrations were analysed. One way for repeated measure analysis of variance (ANOVA) was applied to determine the effect of time and by means of cosinor rythmometry, mesor (mean level), amplitude (half the range of oscillation) and acrophase (Φ, time of peak) were determined. The results showed a seasonal rhythm on Albumin and Alumin/Globulin ratio for sheep and goats, with different acrophases, winter for goats and spring for sheep. A seasonal rhythm was shown also in Alpha 2 globulins by sheep and in Beta globulins by goats. The difference in the acrophase can be attributed to a different production pattern of melatonin in goat, so the liver production of albumin is major during the winter that has a longer scotophase.

A total of 238 Pannon White growing rabbits were slaughtered at 74, 84 and 94 days of age, at an average body weight of 2.53, 2.84 and 3.15 kg, respectively. Within each age group five body weight categories were formed in such a way that the weight difference between two neighbouring groups was equally 0.3 kg. The design of the experiment permitted the separate examination of the effects exerted by age and body weight on carcass traits and on meat chemical composition of the left hind leg (HL) and that of m. Longissimus dorsi (MLD). Dressing out percentage significantly increased with the increase of age in all body weight categories, as well as with the increase of body weight. The effect of age on the ratio of the fore part of the carcass depended on the body weight: significant difference was found in lower body weight categories. The ratio of the intermediate part of the carcass was not affected by age but it increased in rabbits of higher body weight. The ratio of the hind part of the carcass increased in older animals, while it decreased in rabbits of higher body weight, and this decrease was higher in older rabbits. The ratio of the perirenal fat to reference carcass significantly decreased with age in all body weight categories, but it significantly increased in rabbits of higher body weight. With the advancement of age the moisture content of the HL increased, while that of the MLD did not change. With the increase of body weight, moisture content decreased in both the HL and in the MLD. The fat content of both HL and MLD decreased in older animals, while it increased in rabbits of higher body weight.

The aim of the study was to provide information on the genetic variability of the Hungarian Bronze turkey gene reserve population and its difference from the Broad-breasted turkey, and offer guidance and proposals for its future conservation strategies. Altogether, 239 Hungarian Bronze turkeys from 10 strains and 13 Broad-breasted turkeys as a control population were genotyped for 15 microsatellites. All loci were polymorphic with the average number of alleles per locus 3.20 ± 1.146 in the Hungarian Bronze turkey. The mean expected (Hexp) and observed heterozygosity (Hobs) were not different (0.392 and 0.376, respectively) in the overall population, and similar values were obtained for hens and bucks and among hen strains. Inbreeding coefficient (FIS) and Shannon index (I) indicated that there was low inbreeding within hens and bucks. Our results confirm that the genetic diversity in the Hungarian Bronze turkey population has been preserved by the rotational mating system. Differences between the Hungarian Bronze turkey and the Broad-breasted turkey populations were determined. Nei's unbiased values clearly indicated that the two populations are highly genetically differentiated.

The membrane attack complex (MAC) of the complement system is known as a natural immune mechanism of hosts against invading pathogens. This study was undertaken to structurally characterize and computationally analysed the 5′-region, i.e. the downstream promoter region and the 5′-UTR nucleotide sequence of the porcine C8A, one of the components of the MAC. Sequencing approximately 950 bp of the 5′-region and analysing in silico revealed the transcription start site, the TATA-box, the CAAAT-box, and the GATAAbox. High identity was found in range of 37-74 % among the sequences of pig, human, cattle, and mouse. Transcription factor binding sites, such as NFκB, Oct-1, HNF1, CDP, and C/EBP, showed high conservation between vertebrate animal species, especially between human and mouse, or pig and cattle. Seven single nucleotide polymorphisms were detected in the breeds Hampshire, Duroc, German Landrace, Large White, Pietrain, Berlin Miniature Pig, and Muong Khuong. Nucleotide exchanges could cause the generation of new binding motifs, which may affect the expression of the porcine C8A, particularly the C/EBP regulation of the porcine C8A gene-as described in the human C6 and C7 promoter.

Precision Animal Breeding

We breed animals for four principal reasons: as sources of usable products or services; for medical or scientific research; for aesthetic, cultural or ethical considerations; and as pets. The first leads to animal husbandry and livestock breeds of domesticated species kept for food, fibre and other services such as transport and power; the second provides laboratory animals of defined genetic lines, including animals with gene knockouts; the third encompasses breeding for conservation; and the fourth leads to companion animals used for pleasure or recreation. Objectives of sustainability should apply in each case. Precision animal breeding should be used whenever animals are bred *for a purpose*; that is when they are bred for a particular use, environment or market. It should therefore apply when animals are bred for any of the four reasons described above. This is appropriate as the techniques used are largely generic. Our aims in this paper are: (i) to review the methods currently available, (ii) to indicate how they have developed over time to become more precise, (iii) importantly, to show how generic technologies are giving opportunities to substantially improve precision, and (iv) to show what needs to be done to deliver these opportunities. In many cases, the breeding of animals is controlled or influenced by legislation, or by national and international bodies. It is not possible to review precision animal breeding without reference to either of these factors, or the sizes of animal populations or their economic and cultural significance, and for this reason it has been necessary to set the science within a wider background. On the other hand, we have had to limit the range of species considered and will therefore concentrate on terrestrial vertebrates.

The twentieth century marked a turning point in our relationship with other species from which there is no way back. The human population explosion led to widespread competition with other species for agricultural land, and many species became extinct or are now threatened with extinction primarily as a consequence of these activities. We also began to recognize how our industrial culture influences our environment not only locally but also at a distance through atmospheric pollution (e.g. acid rain), and globally through climate change.

We accept that where animals are already within our care, we are responsible for their quality of life. We accept that the activities of man affect all the living things with which we share this planet. But we are slow to realize that as a result we have a duty to care for all living things. This duty extends to the breeding of animals for which we are responsible.

It is in this context that we propose that precision animal breeding should set the following goals.

1. To increase the scope and precision of predictions of the outcomes of breeding decisions (G1).

2. To avoid the introduction and advance of characteristics deleterious to animal well-being or, more generally, the well-being of the species (G2).

3. To manage genetic resources and diversity between and within populations in accordance with the principles set out in the Convention on Biological Diversity.

While these goals will be discussed in more detail later, we shall advance some initial arguments for their form. The first of these goals is clear in the context of precision breeding! However, we have the potential through genomic technology to be more precise in directing the outcomes to be closer to what we desire, both directly in what we select for and indirectly in reducing the chance of unforeseen consequences. The second goal recognizes not only our responsibility for quality of life in populations that are managed, but also some of the ethical conflicts that arise in conservation activities, where an aim for a captive population is to support or re-establish a wild population, and where domestication through adaptation to captivity may be harmful. The third goal recognizes the force of the arguments clearly articulated in the Convention on Biological Diversity, which apply to all our genetic resources both managed and unmanaged.

To meet these goals for precision animal breeding there is a requirement to establish:

1. breeding objectives in the context of the environment in which the progeny are to be kept (R1);

2. a robust parameterized model of the extent of genetic variation, for example heritabilities of traits selected and genetic correlations among significant traits, together with estimates of breeding values (BVs) for significant traits for individual animals (R2);

3. an understanding of gene expression from genotype to phenotype, including the molecular basis of traits (R3).

In practice, for the vast majority of species all this information is seldom available. At best we meet R1 and R2 and a small component of R3 in a limited number of commercially valuable populations. It should be an urgent goal to extend that number as rapidly as possible in order to meet the goals of precision animal breeding, as once desirable population characteristics are lost or conversely deleterious characteristics are introduced, the desired position can be slow and costly to retrieve. Furthermore, in view of the international trade in genetic resources, this need for information must be considered at an international level. There is a continuum between the genetic structures of small, endangered populations and those of widely used livestock species, and in many cases the same issues affect both. For instance, both the Holstein dairy cow and the Arabian oryx (*Oryx leucoryx*) are affected by concerns over rates of inbreeding. In the case of the dairy cow this results from the widespread use of a limited number of desirable bulls through artificial insemination. However, the degree of inbreeding in these two species differs markedly: for the dairy cow, FH-2.5% in the UK to a 1950 base (Kearney *et al.* 2004) whereas for the oryx in 1985 FH–60% to a 1970 base (Mace 1988). Nevertheless, the methods used to trace an animal's evolutionary history may be applied to both rare and widely bred species: for instance microsatellite markers have been used to understand the evolutionary history and current genetic status of Arabian oryx (Marshall *et al.* 1999), South American camelids (Bustamente *et al.* 2003) and the Jersey cow (Chikhi *et al.* 2004). The same procedures are therefore available for application to all managed species, in both small and large populations. Whether it is appropriate to apply them depends on the question being asked and the cost-effectiveness of the procedure, given that there may be other calls on resources.

The State of the Art

The state of the art of animal breeding is based on integrating aspects of many sciences and technologies, with the degree of sophistication varying considerably between the different areas of interest and activities. The key scientific inputs range from genetics (both quantitative and molecular), statistics and computing science, information technology, and the physiology and endocrinology underlying reproduction and fertility. The integration concerns the objectives of the breeding, testing, recording, evaluation, selection and mating of breeding animals.

The objectives of the different areas of breeding activity vary widely. In livestock, well-defined objectives (R1) are necessary for success and will usually evolve over time in relation to societal concerns. Objectives have been for productivity with higher output per unit input but, while these objectives are still important, there has been a notable shift towards tackling issues related to reproduction, well-being and longevity. The ability to tackle the latter objectives is currently constrained in many parts of the world, including the UK, by a combination of poor information on these traits owing to the lack of routine recording coupled to the poor genetic 'signal' within this information. In conservation activities the objective is primarily to increase population size and conserve genetic variation; however, where conservation management includes breeding for release, an important genotype-by-environment interaction arises since breeding in captivity must address the objectives of non-domestication and fitness for wild environments (Gilligan & Frankham 2003).

To illustrate the state of the art we shall describe dairy breeding as a reference and indicate how other sectors differ in technology and operation. The challenge with dairy breeding is that key traits, such as milk production, are sex-and age-limited, and this determines much of the structure of the industry. These sex-and age-limited traits can only be measured in the mature female, reducing the scope for selecting animals on their own performance and the speed at which selection can occur. This is particularly restrictive as it is the male that has the much higher reproductive rate, particularly since advances in reproductive sciences led to the introduction of artificial insemination some 50 years ago. Therefore, improvement systems have developed that are based upon progeny testing in which the genetic merit of candidate bulls is judged by a large number of daughters (100 or more in some cases) being milked on many private farms across the country. Conducting such a test efficiently is a challenge requiring efficient statistical design, the routine gathering of records on performance, and their collation and storage in databases for evaluation.

The estimation of genetic values for the bulls being tested, and for all other animals in the dairy herd, is made using these data and all the other records collected on dairy cows, using best linear unbiased prediction. This statistical technique is based upon detailed genetic models (R2) and accounts for the diverse environments of the thousands of different farms from which the data has been collected. The computational challenge is considerable: predicting BVs for millions of animals across thousands of farms, for hundreds of traits, requires major computational and statistical efforts. Agricultural breeding companies in the pig and poultry sectors are similarly developed operationally, although less reliant on progeny testing, and with the added benefits of the ability to manage the necessary size of population and the scale of recording from within their own resources. Additionally, several breeding companies use sophisticated selection techniques and mating procedures to minimize the loss of genetic variation in the process of generating improvement (Woolliams et al. 2002), a clear step to achieving G3 within commercial livestock. For some sectors, a limited number of genetic markers may contribute information to the selection process; where this is done the most valued markers are those that address traits such as disease or reproduction where other genetic information usually has poor accuracy. Other sectors such as sheep breeding are less well developed in this respect.

For companion animals and equids, the state of the art for breeding is centred upon the use of pedigree records and shows to exhibit examples of the breed and identify desirable parents, together with the occasional use of reproductive technology to increase the reproductive rates of such parents. In equids, there is a move internationally towards more reliable statistical evaluations based on performance and formal testing of young horses. Animals bred for science differ again. The populations of mice that are used are predominantly well-documented inbred strains and crosses derived from them. The use of transgenic technology and mutagenesis to produce novel phenotypes has increased significantly. Limiting the range of species considered here precludes mention of Drosophila, which continues to make a huge contribution to functional genomics, and zebrafish.

Population Genetics

The physical and behavioural characteristics of organisms are determined by the genes they inherit and the environment in which they develop. In terms of physical characteristics, this is widely accepted. Those doubting the importance of genetics in controlling behaviour should consider that behaviours are determined by physical characteristics of neuronal pathways and interactions (e.g. Kendler 2003).

There are approximately 30 000 genes coding for proteins in the nuclear genome of the human being (Southan 2004), which may be taken as a working estimate for other vertebrates. Genes are composed of sequences of nucleotides, and there are differences between individuals in the sequences of nucleotides within a gene known as *polymorphisms*. There are also differences between individuals in the degree to which genes are expressed due to polymorphisms in the regions controlling transcription or in other genes, which control transcription in *trans*. These differences between individuals account for the genetic differences between individuals, which together comprise the *genetic variation* in a population. Where genes determine quantitatively measurable characteristics, these characteristics are known as *quantitative traits*.

During fertilization, an individual receives one copy of each gene (with a caveat over sex-linked genes) in the gamete received from each parent, with the genes passed in discrete blocks called chromosomes. In humans, the number of chromosomes passed from parent to offspring is 23, and in cattle 30. The chromosomes passed by the parent will be a random choice among the pair of homologous chromosomes carried by the parent made during the process of gamete formation, called meiosis. However, the process of meiosis involves *recombination* in which segments of the two homologous chromosomes carried by the parent are exchanged. As a result, new gene sequences may arise. However, there are few recombination events relative to the total number of genes. In addition, variation can arise from new mutations within an individual that change a nucleotide or involve more complex deletions and rearrangements, and change the functional properties of a gene. If these occur within cells of the germ line, these mutations can enter the gene pool of the population and add to the genetic variation observed.

What is observed is the phenotype, which is determined by the sum of the genetic effects and the environmental effects. The latter is the accumulated impact of events in the life history of an individual and, possibly, its parents or even more distant ancestors (e.g. Benton *et al.* 2005). The primary clue for recognizing the existence of genetic variation is the finding of a degree of resemblance between relatives, especially when their development has occurred in different environments.

While we can observe genetic variation within a population, not all of it can be used to produce a change in the population mean; the portion that can be used in this way is called the *additive genetic variation*. It is knowledge of the extent of variation and its different portions that comprise a requirement R2 for precision animal breeding. Knowledge of the precise mutations that underlie genetic variation is unnecessary in order to achieve improvement. For example, despite the doubling of milk yield since 1940 and the many genes that affect milk yield, no single mutation having a direct and verifiable effect on yield was known until 2000 (Andersson & Georges 2004). The better our understanding of the variation, advancing to knowledge of the important mutations underlying variation, i.e. including addressing requirement R3, the more capable we will be of achieving the goals G1 and G2 of precision animal breeding.

The superiority of the offspring from an individual parent when compared with the population mean leads to the concept of the *BV*, which is a measure of the additive component of the parent's genes and its suitability for *selection*. Most quantitative traits are determined by many genes acting in concert (i.e. are *polygenic*), and so an individual's BV cannot be considered the property of a single gene, but of many genes in combination. Breeding schemes are a long way from achieving full precision so man-made selection (*artificial* rather than natural selection) is made on an *estimated breeding value*.

Genetic Correlations

Since some genes affect more than one trait, and since some genes, being located close to one another on the same chromosome, tend to be inherited together, there are *genetic correlations* between certain traits. That is, genetic improvement through selection for one trait may lead to a change (improvement or otherwise) in another. This has been recognized since selective breeding began; Darwin, in *The origin of species by means of natural selection* states: Hairless dogs have imperfect teeth; long-haired and course-haired animals are apt to have, as is asserted, long or many horns; pigeons with feathered feet have skin between their outer toes; pigeons with short beaks have small feet, and those with long beaks large feet. Hence if man goes on selecting, and thus augmenting, any peculiarity, he will almost certainly modify unintentionally other parts of the structure, owing to the mysterious laws of correlation.

Genetic correlations have resulted in undesirable and unexpected side effects of selection in many species. In dairy cattle for instance, selection for yield has resulted in progressive loss of fertility, so that while yield has increased annually in the UK by approximately 90 kg per lactation for the past 20 years, rates of fertility (measured as per cent of animals conceiving to first service) have declined during this period by about 1% per annum (Royal *et al.* 2000): the genetic correlation between yield and calving interval in UK Holsteins is 0.27, and between yield and days to first service, 0.67 (Wall *et al.* 2003). To counter this effect, *fertility indexes* have been introduced initially in Nordic countries and more recently in the UK (Flint *et al.* 2004) to promote those bulls with high yielding, fertile daughters which, though fewer in number, are present in the population. Examples of other genetic correlations that have adversely affected breeding outcomes include the correlation of growth with egg production in poultry (Pirchner 1986) and fleece weight with fertility in sheep (Fogarty 1995). It is only by understanding the mechanisms that lead to genetic correlation, through advances in knowledge of gene expression, that we can be truly proactive in achieving goals G1 and G2, rather than only responding to what is observed.

Molecular Genetics, Genome Sequencing

The revolution in molecular genetics, and particularly genome sequencing, has already provided benefits for animal breeding, but in comparison with what the future holds, our present tools will undoubtedly be seen as primitive. At the time of writing, complete genome sequences are available for a number of species, genome sequences for the chicken, cow, horse, mouse and chimpanzee are either completed or nearing completion, and single nucleotide polymorphism (SNP) libraries for these species are growing rapidly. This information will underpin most of the developments in livestock breeding and breed management during the coming two decades.

Molecular Markers, Microsatellites and Single Nucleotide Polymorphisms

The value of molecular information in assisting breeding decisions has already been demonstrated, in particular through the use of marker-assisted selection as well as for monitoring population structure, and for information on the history and development of populations. Early examples of gene discoveries in livestock that were subsequently used for marker-assisted selection were: a mutation causing malignant hyperthermia in pigs (Fujii *et al.* 1991); a microsatellite closely linked to the mutation causing weaver disease in cattle (Georges *et al.* 1993); a marker for increased litter size in pigs (Rothschild *et al.* 1996); and a deletion leading to double muscling in cattle (Grobet *et al.* 1997). The use of DNA markers for monitoring intra-breed structure has been verifiably illustrated by Chikhi *et al.* (2004) in their study of Jersey cattle on the Isle of Jersey.

The primary route of development for DNA marker technology has been from restriction fragment length polymorphisms (RFLPs; which arise from loss or gain of sites where genomic DNA is cut by restriction enzymes), through microsatellites (arising from variation in the lengths of repeated sequences of satellite DNA) to SNPs (single point variations in nucleotide sequence).

Among other marker types, SNPs share the benefits of being: (i) widespread throughout the genome and so close to coding regions, (ii) co-dominant hence heterozygotes can be identified, and (iii) amenable to the use of PCR to amplify the signal. The use of microsatellites has been widespread over the last decade, with advantages over RFLPs in the multiplicity of alleles at a single locus and a greater repeatability, although reproducibility across laboratories is still a problem.

The technology surrounding SNPs has recently improved by orders of magnitude, so that assays can be scaled and automated, resulting in a low cost that outweighs the disadvantage of being predominantly biallelic. Blott *et al.* (2003) estimated that it is necessary to have five or six closely linked SNPs to replace the information contributed by a typical microsatellite, but given the current approximately 100-fold difference in cost per genotype, the benefit is clear. One outcome of such a reduction in cost is that a cattle genome can be densely marked with 50 000 SNPs at a current (2007) cost of approximately £200; this is a small cost when compared with the cost of breeding and testing a dairy bull for example.

Changes in nucleotide sequences within coding regions are termed 'silent' if they make no change to the amino acid coded, but if silent they may change transcription rates or transcript turnover, thereby affecting protein levels. However, the complexity of control of DNA transcription is still poorly understood. An example of this complexity arising from a single SNP is the polar overdominance in the ovine callipyge gene (Freking *et al.* 2002).

Mitochondrial DNA

Mammalian mitochondria contain a small circular DNA plasmid of 16.5 kb which codes for 37 genes required to be expressed within the inner mitochondrial membrane. The mitochondrial genome evolves 17 times faster than nuclear DNA, probably due to lack of DNA repair mechanisms. As a result, the mitochondrial DNA sequence can be used to monitor evolution on a shorter time scale than is possible with chromosomal DNA. Mitochondrial genome polymorphisms are therefore frequently used to analyse population structure and demographic history. Mitochondrial DNA is haploid, and so each individual has a single haplotype. It is maternally inherited as a result of the limited contribution to the zygote of mitochondria from sperm. This limits its use in relation to domestic species where gene flow through the male line forms an important determinant of evolution and population structure, for instance through artificial insemination, but provides the advantage that it allows introgression through the female line to be distinguished from that through the male.

The value of mitochondrial DNA for animal breeding is that it allows an understanding of population history and structure in time and space. For instance, a low level of mitochondrial DNA polymorphism within a species suggests it has survived a reduction in population size or bottleneck, whereas a high level of variation is characteristic of a large and well-established population. Mitochondrial DNA is extremely valuable in resolving important taxonomic questions when distinguishing subspecies and in identifying evolutionary significant units. Mitochondrial DNA bar coding has been suggested as an aid to assessment of biodiversity (Hebert *et al.* 2003), but this has been questioned, particularly in species subject to parasite infestation or with a high incidence of symbiont infection (e.g. arthropods; Hurst & Jiggins 2005). The general conclusion from large-scale analyses of livestock populations, which are perhaps the most informative for this purpose, suggests that variation in maternal lineages explain at most a small fraction of the variation in traits of commercial interest (e.g. Roughsedge *et al.* 2000a,b).

Epigenetic Effects

Certain heritable characteristics are encoded in DNA by covalent modifications to chromosome structure rather than the nucleotide base sequence of genes. Examples of these modifications are the acetylation of histones in nucleosomes, which alters the ability of the transcriptional enzymes to open and copy the DNA, and the methylation of cytosine residues in DNA, altering their protein-binding characteristics. These alterations in chromosome structure are responsible for genome imprinting, which determines whether a copy of a gene received from the mother or father is transcribed. This process is responsible for genetic diseases such as Huntingdon's chorea through revealing a deleterious mutation that remains uncompensated. Imprinted genes are also involved in the expression of the callipyge mutation (Freking *et al.* 2002) in sheep.

Genetic Evaluation

By knowing BVs, changes can be made in traits without knowing what genes are responsible, and this is what has been exploited throughout domestication: the heritability of characteristics and the similarity of relatives. The development of this process has produced the advanced methods available today, of which BLUP is the epitome for the present.

The introduction of accurate methods for the estimation of BVs has been one of the major successes in precision animal breeding. The need for improved statistical techniques was recognized following the development of artificial insemination during the 1940s, and the introduction of progeny testing for dairy bulls in the 1950s. The method used at that time was *contemporary comparison*, which used production data on a bull's daughters during their first lactation, compared with other cows in the same herd during the same year and season. Effects were then combined across herds.

Contemporary comparison recognized the importance of environmental effects, but failed to take into account differences between herds in genetic merit. It also assumed that all animals other than the daughters of the sire under test were unrelated, and this became increasingly invalid following introduction of artificial insemination. To overcome these difficulties, a method was required which allowed for a wider variety of environmental variables, and took account of the genetic relationships between the animals sampled. These requirements were met by the introduction of BLUP developed by C. R. Henderson in 1949 (Henderson 1975). This statistical method uses matrix algebra to solve the large numbers of simultaneous equations generated for estimating BVs and for identifying environmental effects simultaneously. Implementation of BLUP required considerable computing power to solve the matrices generated. In particular, the complexity of the genetic information used dictates the amount of computer space or time required. In dairy cattle breeding, sire model BLUP evaluations were first used in the US in the early 1970s, sire–maternal grandsire BLUP evaluations were introduced in the UK in 1979 and animal model BLUPs in 1992. BLUP is the current method of choice for evaluation in all sectors and is widely implemented.

Reproductive Technologies

The development of population genetics methodologies and their application to livestock breeding coincided with the development of reproductive physiology and the introduction of reproductive technologies during the last half of the twentieth century. These technologies include artificial insemination and embryo transfer, both of which have been used extensively in the international dissemination of genetic resources. The development of assisted reproduction techniques in human medicine has gone hand-in-hand with their application in animal breeding, advances in human medicine being dependent on studies in animal models.

Artificial Insemination and Multiple Ovulation and Embryo Transfer

Examples of the impact of reproductive technologies on the dissemination of improved genetics are to be found in the turkey and dairy breeding industries. Commercial turkey breeding is now dependent on artificial insemination, because the large size of the males of broad-breasted turkeys, which have been bred for body conformation, precludes natural mating (a failure to apply precision animal breeding). However, artificial insemination provides great benefits. Without semen cryopreservation and artificial insemination the dairy industry would not have developed in the way that it has, since the 1950s, through the international trade in semen and embryos, and these techniques have also had an impact, though less dramatically, on the beef and sheep industries. Since these methods were developed, we now have semen and embryo sexing and multiple ovulation and embryo transfer (MOET) technologies, somatic nuclear transfer and assisted reproduction techniques developed in human medicine, such as intra-cytoplasmic sperm injection. All these techniques, but particularly semen cryopreservation and artificial insemination, play a major role in the design of breeding programmes and in the dissemination of advanced genetics, although the use of nuclear transfer is limited at present. These technologies allow better identification of merit and increased selection intensity through increasing reproductive rate and better use of resources for testing.

The benefits of artificial insemination do not require its widespread use. For instance, artificial insemination makes *sire referencing schemes* possible, where a number of farms cooperate in a breeding programme, using the same reference sires. This has the advantage that by using a small number of semen donors, accurate genetic comparisons across farms can be obtained. These schemes have been extensively used in breeding sheep and beef cattle, and have also been used in the UK for red deer.

Transgenesis

Use of gene insertion and gene knockout techniques has been largely limited to the production of animals for medical research. An exception has been in the production of animals secreting proteins of pharmaceutical value in their milk (Colman 1996). The artificial modification of genes encompasses transfer of genes from one species to another, or altering the genome to remove or suppress the expression of existing genes. Within livestock breeding there are no transgenes within the gene pools of commercial populations at present, outside the experimental stations of various countries. In some countries, for example Norway, the use of transgenesis in livestock is prohibited on principle. What is clear is that the difficulties of ensuring the necessary integration of the transgene into the genome (both in controlling expression of the transgene and integrating expression with the remainder of the genome) has been a challenge of questionable cost–benefit. Early efforts in this area (e.g. Wagner *et al.* 1983) clearly failed in relation to principle G2 that we advocate for precision animal breeding. However, future advances in understanding may lead to applications that satisfy all the principles listed and which will have a benefit sufficiently large to carry public opinion, for example in the prevention of endemic diseases and zoonoses.

Nuclear Transfer and Cloning

Besides its potential to contribute to the study of developmental biology, somatic nuclear transfer has two practical advantages: as a route to transgenic animals and stems cells, and as a means of conserving nuclear genomic material. Cloning by nuclear transfer has now been applied in many species: sheep (Dolly, the first from an adult cell; Wilmut *et al.* 1997); goats; cows; pigs; horses; a rhesus monkey; rats; mice; dogs; cats; rabbits; and gaur.

As a means of producing transgenic animals, the advantage of nuclear transfer lies in the modification and selection of nuclear donor cells in culture, before transfer of their nuclei into enucleated oocytes. Following transfer, factors in the oocyte cytoplasm reprogramme the donor nucleus as a result of which it reverts to totipotency. In this respect, the technique provides the advantages of embryonic stem (ES) cells in species other than mice, which is important because stem cells have not been derived for domestic species. All the gene transfer methods applied to mouse stem cells can be used in cells for nuclear transfer, including random gene insertion, targeted gene replacement by homologous recombination and targeted gene insertion. Selection after transfection can be by toxin resistance, and changes can be monitored by standard molecular biology techniques before cells are used. Despite some improvements in the success rate of nuclear transfer (Campbell *et al.* 2005; Lee & Campbell 2006), much still needs to be done to understand the limitations inherent in the process, and to develop more efficient procedures.

The advantage of nuclear transfer as a means to conservation lies both in the ability to freeze somatic cells and in its use to rapidly increase population sizes. Somatic cells are easier to obtain than gametes, although freezing/thawing regimes are better developed for sperm and eggs in a number of species than for other cells. Once transferred to oocytes, the population of cells represented by the conserved culture is rapidly promulgated. For instance, in cats, large litters can be obtained from the normal mating of individuals, both of whom were produced by nuclear transfer, and this methodology is applicable to rare species such as the African wild cat (Gómez *et al.* 2004).

Applications of nuclear transfer to agriculture include not only the preservation of valuable individuals and rare breeds, but also the rapid propagation of high-genetic-merit animals and animals of specific genotypes (Woolliams & Wilmut 1989). In this respect, nuclear transfer will benefit precision animal breeding by increasing the predictability of livestock performance (G1) and, perhaps paradoxically, fulfilling G3, although Woolliams and Wilmut highlight the risks associated with reducing local diversity through the use of nuclear transfer.

Value Added by Selective Breeding

A conservative estimate of the annual value of livestock production in Europe is €123 billion, and the annual genetic gain at the level of the producer is equivalent to 1.5% of that figure, i.e. €1.8 billion (FABRE-TP 2005). The annual research and development costs of breeding organizations, including collecting data for estimating BVs and carrying out breeding programmes but not product marketing, is approximately €150 million, a benefit to cost ratio of 10. Genetic gains are permanent and cumulative so that the gain made in one year will give benefits over all subsequent years without further intervention, and this increases the benefit to cost ratio cohort by cohort. This contrasts with vaccination strategies, for example, where the benefit from a vaccine requires repeated application, cohort by cohort. The livestock sector in Europe employs 3.5 million people and remains the largest sector in agriculture in terms of both employment and output value. Therefore, genetic progress is central to the success of this major industry.

Structure of the Industry

Progress is principally in the hands of a small number of large institutions. The costs associated with collecting information on economically significant traits, developing and maintaining extensive pedigree and other databases, using up-to-date statistical techniques and large computer systems are high, and as a result much of the expertise is concentrated in a few large commercial companies. For example, over 90% of global poultry breeding stock (layers, broilers and turkeys) is in the hands of two or three organizations selling to worldwide markets in each case. Similar situations apply for pigs and dairy cattle. On the other hand, there are a large number of small breeding organizations dealing with individual breeds, which are unable to benefit easily from the advances open to larger companies. This barrier can be overcome through cooperatives, and effective examples can be found in many European countries (e.g. the Genesis Faraday Partnership scheme in the UK).

Aims for Genetics Within Livestock Industries

The general aim of the livestock sector is to meet the aspirations of the world's population for increased availability of animal products in a sustainable manner while ensuring food safety, animal welfare and the maintenance of rare and specialist breeds. The expected worldwide increase in consumption of animal products for the next decade is 7% annually. The availability of marginal land suitable for producing this increase is limited, and hence there is a need to produce more from the same resources (Food and Agriculture Organization 2000). For genetics to contribute half the required acceleration would mean a doubling in genetic gain. This illustrates both the need and the opportunity for livestock breeding, and placed alongside the requirements for sustainability, the need for precision animal breeding. Sustainability in livestock production implies meeting production targets while ensuring targets are also met for environmentally significant outputs, human feed efficiency, animal health and welfare and maintenance of biodiversity, in both farmed livestock and wild species affected by animal husbandry.

Modern agricultural practices often reduce, rather than increase, genetic diversity in domestic animal populations. Selection for desirable traits and the rapid dissemination of genetic material through populations by artificial insemination and embryo transfer, including international trading in genetic material, all tend to reduce genetic diversity. Furthermore, genetic diversity is lost through the reduction of population sizes in rare breeds, or in some cases through the loss of the breeds themselves. The genetic diversity being lost in this way is not fully characterised. Such trends are contrary to goal G3 and are explored further below. Within breeding schemes goal G3 can be addressed using selection procedures that explicitly manage genetic variation whilst maximising gain through controlling the rate of inbreeding (e.g. Meuwissen 1997).

'Pharming' and Xenotransplantation

Most pharmaceuticals currently available are small molecules. Their chemistry is accessible and they are relatively cheap to produce. In contrast, many therapeutic compounds are proteins, and advances in functional genomics and proteomics will identify many more. Proteins are, however, costly to produce at the purity required for clinical use through current cell culture techniques, and it would be highly desirable to have available alternative methodologies. Secretion of proteins into milk and other biological fluids offers one such route to production of valuable proteins. For example, to bring on-stream a cell culture facility for drug production takes 4 to 5 years and costs over £100 million, whereas to produce a transgenic founder animal and develop a herd of cows derived from it would cost less than £5 million (Forsberg & Bishop 2002) and take no longer. The market for recombinant therapeutic proteins was US$12 billion in 1998, with an estimated 12% growth annually to 2006 (Jasuja 2000). For monoclonal antibodies, the market grew between 1999 and 2001 from $900 million to $3.5 billion, based on only 10 products, with a further 270 products under development.

For these reasons, several companies are in the process of developing products through either microinjection or nuclear transfer in sheep, goats, cattle and pigs. For secretion into milk the most commonly used promoters are derived from the β-lactoglobulin or α-S1-casein genes, and the products include $α_2$-antitrypsin, factor IX and fibrinogen. Other routes for secretion might include blood (human antibodies), urine and seminal plasma (growth hormone), each with different advantages and disadvantages: for example, one advantage of seminal plasma is the existence of the blood-testis barrier, which prevents compounds produced in the testis from acting systemically.

An alternative approach to the therapeutic use of transgenics is the modification of pig organs for transplantation into human patients. Apart from the dangers of viral transmission, this field has been slowed by the need to modify pig tissue antigens. The human immune system rejects pig tissue by an antibody response to the disaccharide galactose-α-1,3-galactose, which is present on the cell surface in porcine tissues but absent in humans. A great deal of effort is currently being put into the development of transgenic pigs lacking this antigenic epitope, and the availability of nuclear transfer in pigs should assist in their production.

The success of these procedures relies heavily on advancing requirement R3 for precision animal breeding, an understanding of gene expression and its contribution to phenotype.

Opportunities for this Century

We are at the threshold of an era where our assumptions of what traits can be addressed by breeding, how merit is assessed and the impact breeding may have will need to be completely revised, primarily due to developments in DNA technology. Although significant strides have been made in applying the principles of precision animal breeding to domestic livestock, there are good reasons to assume that the opportunities for livestock breeding will advance much further in this century. The size of the industry associated with livestock breeding will provide a pull for technical advances, many of which will also be applicable in other sectors.

One reason for optimism is the potential for implementing *genome-wide selection* (Meuwissen *et al.* 2001). Genome-wide selection will use the dense SNP maps emerging from genome sequencing projects. The philosophy of genome-wide selection differs fundamentally from the twentieth century approach of using DNA to enable selection to use marked quantitative trait loci (QTL) or individual genes: genome-wide selection is not concerned with how many QTL there actually are, or where they are located, but rather predicts BVs after weighing uncertainties associated with each small segment of DNA. Dauntingly expensive in the past, the new SNP technologies have reduced the costs of dense genotyping in each animal so that, if not affordable now, then it soon will be. Genome-wide selection tracks gene flow and segregation in all segments of the genome and avoids many of the problems associated with QTL discovery and usage within breeding schemes that has hampered the application of DNA technology to date.

To carry out genome-wide selection, breeders will need to obtain DNA from individuals, as is done now for pedigree testing. A large number of SNPs (tens of thousands) would then be typed in each sample. This vast array of genotypes for animals would be set alongside the recorded phenotypic data on performance, and subjected to an analysis that is complex and computer-intensive, but robust, and analogous to current genetic evaluations. The potential benefits that genome-wide selection offers to livestock breeders are: (i) increased accuracy with minimal cost to inbreeding (Woolliams *et al.* 2002), (ii) ability to overcome age limitations for traits that can only be measured late in life, (iii) the opportunity to overcome or reduce sex limitations, or more generally limitations caused by measuring only special subsets, (iv) use in non-pedigree populations (e.g. identifying desirable DNA fragments in commercial populations that may be selected for within a nucleus), and (v) a direct link between the genetic evaluation and the genome. In some sectors, additional costs might be directly offset through changes in the structure of the breeding programme. These benefits would be expected to drive further innovation in both the structure and goals of breeding schemes.

To exemplify the benefits offered by genome-wide selection in the context of the aims of precision animal breeding, a goal for turkey breeding involves the improvement of growth and the maintenance of egg production. This goal is challenging owing to an antagonistic genetic correlation between these traits (Kranis *et al.* 2006); moreover, the genetic signal is stronger in growth than egg production, and the former is measurable in both the sexes, while egg production is measured only in mature females. For these reasons, traditional indices have a difficulty in making progress in growth without reducing egg production. The potential of genome-wide selection is that the genetic signal for egg production will be stronger in the young males because DNA fragments across the genome can be tracked directly from a turkey hen to her son, making it feasible to deliver the desired goals more precisely.

A further advance towards goal G1 will result from better predictive understanding of genetic correlations arising from fulfilling requirement R3. In particular, genetical genomics with its focus on understanding variation in gene expression will close a gap (the *genotype–phenotype gap*) in biology, linking variation in the genome to variation in flows through metabolic pathways, the fields of the geneticist and physiologist, respectively. Understanding the physiology mediating genetic advance will expand understanding of the physiological consequences. In turn, breeders might anticipate in what way, and to what degree, selection may have adverse consequences, and how these may be avoided.

It is therefore important to recognize that the paucity of data recorded on individual animals limits our horizons. The cliché: 'you can't manage what you do not measure' is directly relevant. Only through the accurate recording of performance and the analysis of records will the opportunities provided by these genetic advances be realized. We need to make better use of data that are already available, by establishing monitoring procedures that generate usable databases, linking these to other databases containing genetic, management or other performance information, and ensuring public databases are open for analysis by scientists. There are examples of good practice in this area such as the British Cattle Movement Service, which makes data available for specific projects. Barriers to such open availability of data need to be overcome where possible, most obviously in ensuring appropriate confidentiality and in promoting data quality when recording on farms where staff time for such activities is restricted.

High-throughput sensor technologies are also developing, mostly based on biospectroscopy, which has the potential to automate many aspects of recording, with increasing miniaturization and portability of the sensing equipment. This will include technology in milking machines for online hormone assays as indicators of fertility and disease, and for monitoring and recording in abattoirs.

Precision animal breeding has much to offer in disease control. Thirty-six farm animal diseases have a genetic basis for susceptibility (Bishop *et al.* 2003), but at present in the UK we do not monitor any of them in genetic databases. The cost of animal disease to the UK livestock industry is £1.7 billion annually, and the impact on production is 17% (between 35 and 50% in developing countries). Some (zoonotic) animal diseases are transmissible to humans. One specific and far-reaching aspiration is to integrate genome-wide selection into routine disease surveillance within the UK. Currently, Defra surveillance provides little genetic information of value, yet given the potential of the approach to dispense with the need for pedigree information, genetic surveillance may be secured by taking DNA-containing samples from casualties paired to a control sample from another animal on the same farm. This will then directly lead to a genetic strategy, delivered through genome-wide selection, coupled to more rapid and direct gene discovery arising from bridging the genotype-to-phenotype gap that is intrinsic to the approach. While the opportunity is extremely promising, research is needed to justify and quantify the benefits in order to persuade decision makers of the value of this approach.

Models for Disease

Spontaneous mutations, particularly in rodents, have led to important advances in functional genomics. Examples are the Brattleboro rat, which provided a model for the study of vasopressin production and function, and *ob/ob* (leptin-deficient) and *db/db* (leptin receptor-deficient) mice, which led to the discovery of leptin and its role in obesity. Since the sequencing of the human and mouse genomes, and the opportunity to understand the roles of genes in disease, the use of mutants as tools for the discovery of gene function has become more widely used. At present, the principal tools used in these studies are strains of rodents (predominantly mice) with characterized genetic mutations. There are in general two methods of generating genetic mutations in rodents for the development of models for investigating gene function: by chemical mutagenesis or targeted transgenesis.

Chemical Mutagenesis

Many laboratories worldwide are generating random mutations in the mouse genome by treatment with chemical mutagens such as the alkylating agent *N*-ethyl-*N*-nitrosourea. This compound causes point mutations in the double-stranded DNA of spermatogonial stem cells through mismatching to an alkylated base. Treated mice have a 1000-fold increase in the rate of mutation. The mutations occur at random in the genomic DNA, and the offspring must be screened for abnormalities in the physiological system or organ of interest. Owing to their random nature, collaborative schemes have arisen to promulgate mutant strains between laboratories, so as to make efficient use of them. Examples of the application of these methods to particular systems are given for defects in the visual system by Thaung *et al.* (2002) and for cardiovascular disease by Svenson *et al.* (2003). At least 1000 strains of mice derived in this manner are currently in use for functional genomics studies.

Transgenesis and Gene Knockouts

Introduction of additional genetic material to the genome can be accomplished through transgenesis. This can be achieved either by pronuclear injection, where foreign DNA is injected into a fertilized egg, by viral transduction (Whitelaw *et al.* 2004) or by transfection into ES cells. Genetic markers can be used to provide for selection of transgenic cells, before transfer to an embryo using a procedure known as *gene trapping* (Skarnes *et al.* 1995).

Alternatively, where a gene sequence is available and a targeted deletion is required in order to study the gene function, genes can be selectively removed. This is achieved by homologous recombination with a mutated version of the gene in a synthetic construct, usually in mouse ES cells, but also potentially in somatic cells for nuclear transfer. ES cells are then introduced into a conceptus and adult mice derived from them, which are bred to produce offspring homozygous for the desired deletion. Co-transfected markers such as drug resistance genes can be used to allow selection of appropriate cells, and the deletion can be verified in offspring by Southern blotting. Heterozygotes and wild-type mice of the same strain are available as controls. The phenotype of the knockout progeny reveals the gene's normal role.

However, in some cases genetic models produced in mice are not closely applicable to human pathology, as gene functions differ between the species. As a result, other species are being investigated. For instance, the human cystic fibrosis mutation does not lead to a comparably severe condition in mice, and sheep may provide a more suitable model (Harris 1997). In addition, there is a project to create a pig model of neurodegenerative disease, culminating in the birth of pigs lacking functional copies of the ataxia-telangiectasia mutant (A-TM) gene. These pigs will provide an alternative animal model in which ataxia-telangiectasia (A-T) can be studied, which the current rodent models fail to reproduce. However, the techniques for producing such targeted mutations are time consuming and unpredictable, since the inability to derive ES cells that contribute to the germ line from domestic species necessitates the use of somatic nuclear transfer. An alternative to the mouse as an animal in which gene function can be studied is the zebrafish (*Danio rerio*). The advantages of this species are the large numbers of accessible, transparent embryos which can accelerate phenotype screening, and as a result the zebrafish is a productive model system for the genetic analysis of embryogenesis, organ development and related pathologies. Over 2000 mutations are currently available in more than 600 genes, and there is a highly developed international collaborative network in place for dissemination of these resources.

Opportunities for this Century

The deliberate production of animals with genetic disorders for research purposes clearly fails to meet goal G2 of precision animal breeding, and random mutagenesis with subsequent identification of effects on phenotype fails to meet G1, which calls for precision in predicting breeding outcomes. For these reasons, it is appropriate that breeding under these conditions should be subject to legislation controlling the use of animals for research; in the UK, this legislation is contained within the Animals (Scientific Procedures) Act 1986. Breeding animals with defined disorders for research purposes meets the objective of the refinement, reduction and replacement (3Rs) enshrined within the Act, since more precise models of genetic disease should lead to refinement and ultimately a reduction in the numbers of experiments performed, through generation of more accurate data.

An important advance in this field will be the isolation of ES cells from livestock species. This will permit the development of animal models for human disease in a way not currently possible. However, it should be noted that although research in this area has been underway for the past 20 years, and despite the generation of chimeras, to date germ line transmission of ES cell genomic information has not been achieved in livestock. Clearly, there needs to be more research on the gene expression patterns characteristic of ES cells. While ES cells remain unavailable for livestock species and somatic cell nuclear transfer is relatively inefficient, enhancing the efficiency of targeted homologous recombination will be an important area for the future.

Conservation Breeding

Breeding animals for conservation differs from breeding for production purposes in three principal ways: the breeding aims are not focused on improving quantitative traits, but are based on population size and genetic variation; the numbers of individuals in endangered populations are by definition low; and stochastic environmental factors play a huge role in determining population viability. Conservation of endangered species and breeds has historically been the concern of two separate scientific communities, those involved in conservation of rare breeds of domestic animals and those working with wild animals.

Rare Breeds of Agricultural Livestock

About 40 species of domesticated mammals and birds are kept worldwide, although only six mammals (cattle, buffalo, sheep, goats, pigs and horses) and four birds (chickens, ducks, geese and turkeys) are widespread. Among these, approximately 7000 breeds have been produced since the beginning of domestication. In the UK, 70 breeds are recognized by the Rare Breeds Survival Trust (20 became extinct between 1900 and 1973, before the Trust was formed). The criteria for 'Rare Breed' status include a numerical basis, current trend in population size and a consideration of the period for which the population has been closed. Efforts have been made to identify and prioritize breeds for conservation action as requiring either maintenance *in situ* or *ex situ*, or cryoconservation (a form of maintenance *ex situ*). Eding & Meuwissen (2001) explored a centrally organized approach to cryoconservation in gene banks, attempting to define objectively: (i) which breeds would or would not be stored, (ii) how many individuals to sample from each breed, and (iii) which individuals should be sampled so as to maximize the genetic variation that can be recreated from a gene bank. This method depended upon deriving relationships among breeds based upon DNA markers. Other related approaches use genetic distances (again based on markers) to define alternative measures of diversity rather than the concept of stored genetic variation that was used by Eding and Meuwissen. Weitzman (1992) provided rules for defining what may or may not be a legitimate utilitarian measure of diversity. D'Arnoldi *et al.* (1998) and Caballero & Toro (2002) provide interesting discussions on these rules. In particular, Caballero & Toro (2002) point out that diversity in a species is concerned with diversity within breeds as well as between breeds, and that this aspect is neglected in methods based on genetic distance between breeds. However, as Woolliams (2004) points out, action is required based upon the best information available at the time, using the best practice that is feasible at the time and involving stakeholders at all stages.

Endangered Wild Animals

The principal threat to wild animals arises from human population growth and the aspirations of the human population for an improved quality of life. The human population, now (2006) 6.5 billion, increased exponentially from an historic base of approximately 300 million at the time of Christ to about 800 million by 1800, and increased fourfold during the twentieth century alone. It is expected to reach 9 billion by 2050. The growth in human industry (economic activity) is even more revealing than the increase in population. During the twentieth century, a fourfold increase in the population was accompanied by an 18-fold increase in the economic output. During that time, we lost about 40% of the planet's forests and 10% of the coral reefs (both major repositories of biodiversity) with additional threats to mangrove swamps, wetlands and savannah.

Extinction is of course a normal part of the life process; for example, since birds first appeared 130 million years ago, 500 000 species are said to have existed, the maximum number extant at any one time being 11 500. Today there are 9946 species of birds, about 1200 of which are threatened (listed as critical, endangered or vulnerable by the Species Survival Commission Red Data List). One hundred species have become extinct during the last 100 years. However, these statistics hide the fact that the rate of extinction, and the level of threat, is about 100-fold higher today than it has been historically, for all groups of vertebrates. For mammals, 1100 of 4763 species are threatened; for reptiles, amphibians and fishes the situation is less well known and the proportions threatened are underestimated (namely 253 out of 7970 species, 124 out of 4950 and 734 out of 25 000, respectively). The evaluation of the extinction threat to a species is an imprecise science, but has been put on a more quantitative basis by the development of criteria for endangerment by Mace & Lande (1991). The maintenance of genetic diversity is one of the goals (G3)a of precision animal breeding.

Ex Situ Versus in Situ Breeding

In making breeding decisions for endangered species it is clearly advantageous to ensure the survival of animals in their home range. However, where this is not possible it is better to remove individuals to an alternative, safer, environment than to allow them to be hunted to extinction. Alternative locations include zoos or wildlife parks, and may be found in a country far from the animal's original habitat. This potentially gives rise to conflicts of interest. Removing individual animals from their home range reduces the number *in situ* thereby putting further pressure on those remaining. Where the *ex situ* site is far from the region or country of origin, there may be political pressures against removal. Local extinction following removal may result in the original habitat being irrevocably lost as a result of the lack of pressure to maintain the population locally, making reintroduction difficult. Clearly where possible it is preferable to achieve a combination of *in situ* and *ex situ* approaches.

Conservation Organizations and Structures

There are two broad categories of organization involved in the management of populations of wild animals: those based in zoos, aquaria and wildlife parks, and those overseen by the United Nations. In general, those representing the UN oversee information on a global scale, collecting and managing information with a view to developing priorities for habitat and species conservation both *in situ* and *ex situ*. Zoos, aquaria and wildlife parks principally oversee conservation breeding and management *ex situ*, but are also becoming increasingly involved in *in situ* conservation. The tools used centre on stud books, which is where the pedigree and other information is recorded, and software packages to manage the stud books.

Four United Nations bodies are involved in different aspects of conservation of biodiversity: the World Conservation Union (WCU; formally the International Union for the Conservation of Nature and Natural Resources, IUCN); the United Nations Environment Programme (UNEP); the United Nations Development Programme; and the Food and Agriculture Organization (FAO). The WCU (IUCN) established Commissions on Ecosystems Management, Education and Communication, Environmental, Economic and Social Policy, Environmental Law, the World Commission on Protected Areas and the Species Survival Commission. The last of these, the SSC, is the largest, with 120 specialist groups and task forces to identify threats to groups of taxa and recommend conservation priorities and actions. These specialist groups include the Conservation Breeding Specialist Group (CBSG) which, working through Conservation Assessment and Management Plans and Global Captive Action Plans, identifies species in need of propagation in captivity with a view to subsequent release.

The principal role of the CBSG is to facilitate decision making by arranging meetings of appropriate experts. By convening and hosting meetings of interested parties, this body develops Habitat

Management Plans for individual species (population and habitat viability assessments, PHVA) or groups and geographical regions (Conservation Action Management Plans, CAMPs). PHVAs use software developed for the purpose, which applies population dynamics techniques, and takes into account demographic events at an individual animal level. The CAMPs use databases to formulate recommendations on actions required to manage populations or habitats. The level applied is wider than in the case of PHVAs, as CAMPs deal with broad taxonomic groups or geographical regions. Although their objectives appear to overlap, the structures organized by the zoo associations run in parallel with CBSG, through common membership; the same personnel are frequently involved in both sets of bodies.

Through UNEP, the World Commission on Environment and Development (the Bruntland Commission) in 1987 highlighted the importance of biodiversity and sustainability, initiating a process of international discussion culminating in the United Nations Conference on Environment and Development in Rio de Janeiro in 1992. This conference led to the Convention on Biological Diversity, which has now been ratified in 188 countries. The three objectives of the Convention are the conservation of biodiversity, the sustainable use of its components and the equitable sharing between societies of benefits arising from the exploitation of genetic resources. In 2002, the parties to the Convention adopted a strategic plan, focusing on reducing the rate of loss of biodiversity globally, regionally and locally by 2010. The seven strategies adopted as means to achieving the objectives of the Convention include: (i) 'Reducing the rate of loss of the components of biodiversity, including: (a) biomes, habitats and ecosystems, (b) species and populations, and (c) genetic diversity' and (ii) 'Promoting sustainable use of biodiversity'.

Notable advances have been made during the past 20 years in the management of endangered animals in captive populations held by zoos and wildlife parks. Stud books are held by individuals, who are most frequently animal managers and keepers in zoos. Permission to hold a stud book is agreed with the local zoo association (the European Association of Zoos and Aquaria (EAZA) in Europe or the American Zoo and Aquarium Association (AZA) in the USA), and ratified by the AZA Wildlife Conservation and Management Committee (WCMC) and the World Association of Zoos and Aquaria (WAZA). Stud books are managed using an internationally accepted software package, which includes compatible software for keeping animal records and medical records. In North America, the AZA represents more than 200 zoos and operates for conservation breeding purposes through the WCMC. In turn the WCMC organizes Taxon Advisory Groups (TAGs, of which there are 46), which are composed of experts on particular groups of animals (taxa), recommending breeding actions to AZA institutions, evaluating the need for captive breeding and assessing the space available in zoos. Their members have expertise in taxonomy, assisted breeding, contraception and wild populations and habitats, as well as educational and training programmes. In North America TAGs develop two kinds of breeding plan, Species Survival Plans (SSPs) and Population Management Plans (PMPs). SSPs (of which there are currently 107, representing 161 species) include genetic and demographic analyses of captive populations, and make individual breeding recommendations for both *in situ* and *ex situ* populations. PMPs (of which there are 282) fulfil the same function as SSPs, but for the *ex situ* population only. Both SSPs and PMPs work through stud books (of which there are more than 400 in North America). TAGs also develop action plans with priorities for *in situ* conservation requirements. The SSP programme started in 1981, and PMPs in 1994.

A similar structure operates outside North America. In Europe, 200 zoos in 25 countries are involved in breeding programmes. The equivalent of SSPs and PMPs are European Endangered Species Programmes (EEPs, of which there are 151) and European Stud Books (ESBs; 140), and these are run through Taxon Advisory Groups (*n*=40) reporting to the EAZA. Similar structures are in place in Africa, Australasia and Asia.

Conservation strategies for zoos worldwide are brought together by WAZA. The 2005 second edition of *Building a future for wildlife: the WAZA conservation strategy*, published by WAZA, provides a blueprint for the roles zoos and aquaria play in conservation of wildlife and their ecosystems. It represents a common philosophy for zoos and aquariums worldwide and defines the policies required to achieve their conservation goals.

Conservation Breeding and Release

There have been notable successes for conservation breeding, which have shown that with enthusiasm, commitment and an ability to collaborate, it is possible to ensure the survival of species that would otherwise have been lost. A good example is that of the Arabian oryx (*Oryx leucoryx*), a gazelle hunted to extinction in its natural range (the southern Arabian peninsula, Aden, Yemen and Oman) in 1972. A survival plan was put in place, largely initiated by the Fauna Preservation Society of the UK (now Fauna and Flora International), between 1961 and 1982 which, with 3 animals wild-caught in Aden in 1962 and 16 others from zoos taken to Pheonix, Arizona, bred a herd of 35 by 1972 and 106 by 1977. Offspring were returned to the Middle East in 1978 and to their original range in Oman in 1982. The cultural benefit of this reintroduction programme to the inhabitants of that region is hard to overestimate, and their involvement in the animals' care is evidence of the important place this oryx occupies in the culture of the desert-living nomads of that area.

Other successes of conservation breeding have been the scimitar-horned oryx, Pére David's deer, Przewalski's horse, the black-footed ferret, the Mauritius kestrel, the Puerto Rican parrot, the black and white ruffed lemur, the Egyptian tortoise, the Partula snail and the Wartbiter cricket. In each case the principles of precision animal breeding have been applied, because the animals involved were bred for a purpose and with limitation of loss of genetic variation (G3) in mind. In some cases the pedigrees of the animals used were not well known, but nonetheless, the principles were applied.

In the case of animals bred for subsequent release from captivity, there is a need for the individual animals involved to be isolated from domesticating influences (Woodworth *et al.* 2002), and to be trained in behaviours required for an unsupported existence. Will an animal raised in captivity be able to hunt, or to identify predators or poisonous plants? Examples of species in which these processes are important include most notably primates, for instance the golden lion tamarin, which has been released into its former range in Brazil.

There is also a need to carefully consider the genetic composition of the released population and its impact on the animals remaining in the captive group. Maximizing genetic diversity in the released animals may deleteriously affect that of the captive group, which should be avoided if further releases are planned, and the captive population is to be maintained. Releasing individuals over-represented by progeny benefits the captive subpopulation, because in a captive breeding programme these animals would not be bred, and their reintroduction will free spaces for further growth of the captive population. Where there is uncertainty over the survival of reintroduced animals this strategy is appropriate, because it protects the genetic diversity of the remaining captive animals. On the other hand this strategy may not maximize the survival chances of the reintroduced group, because that benefits from maximizing genetic diversity in the reintroduced animals. These trade-offs have been modelled for four captive-bred species with different breeding histories by Earnhardt (1999).

Tissue Banking and Conservation by Nuclear Transfer

Gamete storage techniques (sperm and oocyte freezing; germplasm preservation) have given rise to the possibility of the 'frozen zoo'. Recently this has been expanded by the opportunity to propagate nuclear DNA through somatic nuclear transfer (e.g. Gómez *et al.* 2004), offering the possibility to store tissue in the form of cell cultures, which can be frozen, for subsequent nuclear transfer to an enucleated oocyte, through which cloning can benefit diversity (Woolliams & Wilmut 1999). A number of issues arise with these techniques. Firstly, it is relatively easy to freeze gametes, but more difficult to ensure they are fertile on thawing. Secondly, the numbers of endangered animals to which gametes can be transferred are limited, and this limits the information available on processes such as superovulation and induction of oestrus, which are required as part of the technique. Thirdly, somatic cells in culture may undergo chromosomal reorganizations rendering them unfit for nuclear transfer. In somatic nuclear transfer, mitochondria are transferred from the donor somatic cell to the recipient oocyte, and the offspring may be mitochondrial chimaeras; this is particularly likely if a surrogate oocyte (from a related but non-endangered species) is used as recipient. Lastly, which individual animals to preserve? There is an argument to conserve the broadest set of alleles possible from a population; appropriate animals can be identified by genotyping, by measuring a set of quantitative traits or on the basis of coefficients of relationship (Lamberson *et al.* 2002). However, a random stratified sample of the population representing the gene flows in the population as identified by pedigree, performance and molecular analysis may be least at risk from deleterious genes.

The storage in frozen form of genetic material from animals is therefore much more problematic than for plants. But despite the difficulties, it does offer great opportunities, and several centres are engaged in establishing such tissue banks (including in the UK the Institute of Zoology, London and the University of Nottingham).

Opportunities for this Century

One of the key questions arising in making conservation decisions is the identity of the *ESU*, the 'population unit meriting special management' (Ryder 1986). This is defined as 'a set of populations that is morphologically and genetically distinct from other similar populations or a set of populations with a distinct evolutionary history'. This may be smaller than the species, and may not necessarily represent a reproductively isolated population: it may be based on genetic, phenotypic or behavioural criteria. The US Endangered Species Act recognizes ESUs as the unit requiring protection. The identification of the ESU for a particular species most frequently depends upon molecular genetic information, such as mitochondrial DNA and microsatellite polymorphisms. Thus, the application of molecular genetics techniques to conservation questions is of paramount importance. Examples of ESUs identified in this way are the dusky seaside sparrow and the red wolf (although in each of these cases the evidence has been challenged). An example of a species found not to represent an ESU is the Cape Verde kite (Johnson *et al.* 2005) though this study did identify two other species of kite as phylogenetically distinct, the yellow-billed kites from South Africa and Madagascar and those of northern Africa. More needs to be done to define how ESUs are determined for each species, and to apply these techniques more widely.

The developments in requirement R3 that will arise during this century will help in the recognition of selection pressures that lead to adaptation to captivity for conserved wild species. The availability of DNA information will play an increasing role in the management of diversity within an ESU (after having defined it); however it is important to recognize that the scope for managing diversity beyond best practice using pedigrees is limited. Goal G3 of precision animal breeding can be considerably advanced by more informed rural planning to avoid wild populations being fragmented into small isolated groups.

In coping with climate change and the increasing need for conservation work that will arise from it if present predictions are even close to reality, there will be a need for committed personnel, highly productive systems and structures, and well-defined objectives. Above all there will be a need for more 'spaces' for endangered animals in conservation breeding programmes.

Companion Animals, Pets and Animals for Recreation

Large numbers of animals are bred as pets, most frequently by professional breeders. The number bred by individuals for their own use is small by comparison. The range of species bred for this purpose is widening continually, and it is now possible to buy as pets, wallabies, miniature horses and donkeys, and a wide range of other vertebrates. Most pets, however, are dogs or cats; the size of these populations is an indication of their potential to cause mischief: there are 6.7 million dogs and 7.5 million domestic cats in the UK. There is also a feral cat population of 800 000, and each year about 135 000 dogs stray, of which about 10% are put down. There are approximately 975 000 equids (not exclusively horses). For each of these species there are examples of successes and failures in the application of the principles of precision animal breeding, sometimes with highly unfavourable genetic consequences. Online Mendelian Inheritance in Animals lists more inherited disorders in dogs than any other species excluding humans and mice (479, of which only 47 have been characterized at a molecular level), with cattle second (366, 31 characterized) and cats third (273, 13 characterized). In many cases these are diseases with human homologues. Nonetheless, significant advances are now being made in the recognition of problems associated with certain breeds, and in structures which should lead to more widespread use of appropriate techniques.

Cats and Dogs

Selection in the breeding of cats and dogs is based on satisfying human aesthetic criteria rather than those ensuring fitness in a wild environment. As a result deleterious mutations have become widespread in populations, through inbreeding used to 'fix' desirable traits).

In the UK the breeding of dogs is controlled by the Breeding of Dogs Acts of 1973 and 1992, and the Breeding and Sale of Dogs (Welfare) Act 1999. This legislation licences puppy farms and ensures that the dogs are suitably accommodated, fed, exercised and protected from disease and fire. The 1999 Act provides that bitches are not mated until they are 1-year-old, that they have no more than one litter per year and that they give birth to no more than six litters in a lifetime. The Act also requires that accurate breeding records are maintained. This legislation, together with a range of other legislation relating to the welfare of farmed and non-farmed animals, is being brought together under a new bill (the Animal Welfare Bill) presently before the UK parliament. In addition to the above legislation providing for breeding of puppies, breeding of certain types of dog for fighting is prohibited under the Dangerous Dogs Act 1991, which 'prohibits persons from having in their possession or custody dogs belonging to types bred for fighting; imposes restrictions in respect of such dogs pending the coming into force of the prohibition; enables restrictions to be imposed in relation to other types of dog which present a serious danger to the public, and makes further provision for securing that dogs are kept under proper control'.

None of this legislation covers selection in breeding, however, and as described above, many breeds now suffer from significant levels of inherited disease. The ways in which Kennel Clubs and dog shows have operated in the past have led to their rapid dissemination. Therefore, for example: (i) a number of breeds have been established with too few founders, (ii) stud books are closed for some breeds, preventing introgression of new genetic material and leading to inbreeding, (iii) selection has often been based on inappropriate type criteria, and (iv) even if the founders were sufficiently diverse genetically, clubs have frequently failed to exert adequate management of the genetic diversity generation by generation.

Kennel Clubs are generally aware of these problems. Stud books are becoming 'less closed' and phenotypic screening tests are becoming available and in use for many common inherited disorders. However, these do not identify individuals carrying recessive genes or late onset disorders which can only be diagnosed after the animal has bred. Of greater potential value are molecular markers for disease, which are currently being developed for a wide range of conditions. The development of these DNA-based tests has been facilitated by sequencing of the dog and cat genomes.

These tests will identify carriers of recessive diseases such as copper toxicosis and progressive retinal atrophy, which affects several breeds, as well as mutations responsible for von Willebrand's disease in Shetland sheepdogs, Scottish terriers and Doberman Pinschers (where different mutations are responsible for the condition), phosphofructokinase deficiency syndrome in English Springer and American Cocker spaniels and pyruvate kinase deficiency in Basenjis. These tests require small blood samples and can be carried out on prepubertal animals. DNA tests for progressive retinal atrophy in Irish Setters, copper toxicosis in Bedlington terriers, and hereditary cataract and L-2-hydroxyglutaric aciduria in Staffordshire Bull terriers are already available in the UK through the Animal Health Trust and the Kennel Club. In order to speed eradication of progressive retinal atrophy in Irish Setters, there is a ban on breeding and showing carriers for the condition, which can be identified using the test. This is an example of what is achievable in this field, and it is to be hoped that other tests will be introduced with similar openness of information.

In some countries, these tests are required by law. The Netherlands and Germany have legislation banning breeding of animals with severe defects or breeding that could result in offspring with inherited disease. This is a strong implementation of goal G2. In Sweden it has been proposed to limit the number of litters per parent based on the population size for each breed, which will limit the impact of deleterious genes on the population.

There are nonetheless notable instances of excellence in breeding of dogs in the UK. An example of an organization taking a highly professional approach to breeding working companion animals is the Guide Dogs for the Blind Association. This charity breeds over 1000 guide dog puppies annually, and supports about 5000 guide dogs in use in the UK. About £10 per day is invested in the breeding, training and care of a guide dog, which is a measure of the level of investment that should be applied in breeding working dogs.

In contrast to the controls on dog breeding, the breeding of cats is uncontrolled, presumably because cats live more independent lives. Molecular approaches are less well developed than for dogs, but the mapping of the feline genome has recently been reported by Murphy *et al.* (2000), and DNA tests are being introduced (for instance the scheme for polycystic kidney disease run by the Animal Health Trust and the Langford Feline Diagnostic Service).

Horses

Too little attention has been paid to selection in breeding sport horses. It has been suggested for some time (Cunningham 1989) that speed of racing in thoroughbreds has not improved for some years, possibly through lack of genetic variation, or from attainment of a physiological 'ceiling'. More positively, a recent study (Mota *et al.* 2002) has indicated a small genetic trend in racing time improvement, suggesting that failure to observe any phenotypic improvement may be due to only weak selection pressure since selection has been largely on prize money rather than race times. However, such trends need to be better established.

The Jockey Club rules governing thoroughbred breeding preclude artificial insemination or other assisted reproduction techniques, and this limits genetic improvement but reduces the potential to erode levels of genetic diversity. However, these rules do not apply to other breeds, and there is in particular much to be gained from a systematic approach to breeding horses and ponies for sports such as eventing and polo. This has traditionally been carried out with little or no consideration of performance testing, and as a result opportunities have been missed in international sports.

Opportunities are now emerging, however, through legislation affecting the management of horses. Following the publication by Defra of the Animal Health and Welfare Strategy for the UK in 2004, leading to the Animal Welfare bill, and as a result of concerns for the horse industry in England following the introduction of the Act prohibiting hunting with hounds, there have been two strategy documents which offer hope for the future. These reports are: *Joint Research on the Horse Industry in UK*, produced by the British Horse Industry Confederation and Defra, and the *Health and Welfare Strategy for the Horse* produced by the British Veterinary Association and Defra. As a result, there is recognition that opportunities are being lost through the fragmented nature of the UK horse industry in terms of selective breeding, and this has been addressed for eventers by the production of genetic evaluations by the British Equestrian Federation through its breeding arm, British Breeding (Kearsley *et al.* 2006). Secondly, a passport system has been introduced, which will lead to the better identification of animals, so discouraging the indiscriminate breeding of horses and ponies of low quality or genetic merit. One requirement of the passport system is that the document should accompany the horse when it is used for breeding, thereby ensuring correct identification of the animal at that time. The legislation requires the establishment of a National Equine Database from information on passport applications, which will be a valuable tool for improvement in breeding, and, as a result, in the health and performance of competition horses and ponies in the UK. It has also been agreed that artificial insemination can be carried out safely by appropriately trained personnel, and an Exemption Order under the Veterinary Surgeons Act 1966 is currently under consideration to allow this procedure to be performed by people other than veterinary surgeons. These innovations should go some way towards redressing the historical lack of professionalism on the issue of breeding horses for sports in the UK, and help to provide structures which are taken for granted in some other European countries (e.g. Germany). They will, however, have little impact on hill breeds where animals run as herds and the assumption is that the alpha male is the sire of the foals. Although there is some indication that the herds do not mix, parentage cannot be assigned accurately on this basis. Molecular markers have not in the past been widely used in equine breeding, but work is currently underway to map the equine genome. Supported by the Horserace Betting Levy Board and the Childwick Trust an international collaboration (including in the UK the Animal Health Trust and the Royal Veterinary College) has been established aiming to produce a low resolution genetic map for the horse. The current genetic linkage map comprises approximately 800 markers and, excitingly, the horse genome is currently being sequenced. This development should accelerate research progress in horse breeding and permit the development of genetic screening tests for inherited diseases and their use to breed healthier horses and ponies. For example, the molecular defects underlying some of the important inherited diseases in horses (hyperkalaemic periodic paralysis, severe combined immunodeficiency syndrome, lethal white foal syndrome and junctional epidermolysis bullosa) have been identified. Severe combined immunodeficiency affects 8% of Arabian horses (Ding *et al.* 2002) and a diagnostic test which will identify carrier individuals is available. An improved cytogenetic test based on chromosome painting for the X0 sex chromosome abnormality, which renders affected mares sterile, has also recently been developed, which will improve the identification of these individuals. Note added in proof: there is now a complete sequence for the horse genome.

Future Needs in Animal Breeding and Genetics

Genetic improvement in livestock has a truly amazing history, with the *Journal of Animal Science* having been the world's premier publication venue for many of the major scientific innovations that have occurred since 1908 (Hohenboken, 2009). Although the majority of advancements in the past 100 yr have been made using quantitative genetics approaches, in recent years it has been nearly impossible to miss seemingly daily news reports about exciting discoveries in the relatively new field of molecular genetics and genomics. While most of these reports have focused on the unraveling of the human genome and its implications for human health, there has been significant spillover into plant and animal agriculture as well. At times over the past 15 yr, it has seemed that this new and exciting field would hold all of the immediate answers to breeding improved beef and dairy cattle, swine, sheep, and horses. Today we have draft genome sequence assemblies of the bovine and equine genomes completed and made publicly available along with a host of rapidly developing diagnostic tools, with the same now underway for the porcine and ovine species. Such transformational technological developments, coupled with several major societal shifts that are predicted to result in changes of epic proportions for the livestock industry and its producers, make the centennial anniversary of American Society of Animal Science (ASAS) an ideal time to consider needs in the field of animal breeding and genetics to fully address these challenges.

Historical Perspective—Genetic Theory to Selection and Mating System Tools

The first half of the 20th century was an immensely prolific time in agricultural science. Arguably, the most dramatic discoveries were actually in the fields of genetics and statistics. During the 1920s and 1930s, the field of population genetics came of age, primarily as a means of quantifying and describing the writings of Darwin from the late 1800s. The emerging leaders of this field helped to describe the concepts of genes, gene loci, chromosomes, and cellular reproduction. They were also instrumental in establishing the field of biometrics—statistics as applied to biological phenomena. These early statisticians developed much of the underlying theory used broadly in science today. At the same time there were pioneering scientists who had the foresight to develop populations of livestock upon which they began to practice long-term selection and inbreeding.

Scientists also made what seemed to be an unrelated but extremely valuable discovery in plant genetics during this same period with the observation that when 2 unrelated lines of germplasm were crossed (hybridized), the resulting crossbred progeny had better performance than the expected average of the parents. The concept of heterosis between mildly inbred lines was born—and with it the seed industry and crop agriculture was revolutionized. At the time, livestock breeders did not see any great benefit from this phenomenon, but that would dramatically change later (Cundiff, 1970; Dickerson, 1973).

The era following World War II was a particularly exciting time for livestock genetic improvement, as it was in many fields. The 1940s saw some of the greatest minds to ever grace the study of livestock genetic improvement at their prime. J. L. Lush, who many refer to as the modern-day father of animal breeding (Dickerson, 1973; Dickerson and Willham, 1983), was busy defining, with coworkers Hazel and Dickerson, the concept of the "selection index" and "breeding value"(Hazel, 1943; Lush, 1947). The field of biometrics had matured to the point where it was now possible to determine from experimental populations that performance for traits affecting production could be measured, and that many of these traits appeared to be heritable.

In 1953, Watson and Crick presented for the first time in the scientific literature the molecular structure of the genetic code—DNA (Watson and Crick, 1953). Combined with the theories of genes and heritable variation of traits, it was now possible to visualize how these genetic differences at the gene level might one day be exploited for genetic improvement.

Also, in the 1950s, 2 significant events occurred that would permanently change the nature of livestock breeding. The first was that AI techniques matured to an adoptable level for cattle breeders, especially dairy producers. Coupled with the earlier institution of the Dairy Herd Improvement programs of USDA, volumes of data began to accumulate matching pedigrees to milk production records. At the same time, computing technology was beginning to surface as a usable tool, even though it was rudimentary to what we now have today. Dairy cattle breeders had enough foresight, however, to understand the power of coupling quantitative genetics theory to AI; as a result, genetic evaluation as an applied science was born. Furthermore, the plans of the dairy industry did not go unnoticed by other livestock breeding communities, and performance data recording quickly became mainstream practice in all of the species under various models, often at the level of breed registries and societies. Several organizations were also established in the industries to develop uniform guidelines for performance recording, including the Beef Improvement Federation, National Swine Improvement Federation, and the National Association of Animal Breeders. These organizations were provided critical leadership by animal breeders who were leaders in the field and in ASAS.

The late 1960s and early 1970s was the next period of great change in livestock breeding and genetics. Two things occurred somewhat simultaneously: the importation of germplasm from several lines and breeds and the next generation of computing technology coming of age. As a result of their faster growth rates, size, and altered body composition, several of these breeds quickly took a strong foothold in the seedstock industry. Breed associations and initial breeding companies took advantage of developing computer technology in the 1970s to deploy the first across-herd genetic evaluations using sire-maternal grandsire models, and the initial era of breeding value estimation across large-scale populations was born. These efforts built upon selection index theory and its applications employing mixed-model equations, first suggested by Henderson (1959, 1973, 1975) and further deployed by Willham (1979) among many others.

Additionally, producers discovered that hybrid vigour was indeed possible—and very economically beneficial—when many of the new breed lines were used in systematic crossbreeding programs (Dickerson, 1970, 1973). One of the most widely noted efforts in this area was developed and carried out at the USDA-ARS US Meat Animal Research Centre (MARC) through programs like the Germ Plasm Evaluation program led by Larry Cundiff and the Germ Plasm Utilization program led by Keith Gregory. Since 1968, this collective effort has produced the fundamental body of knowledge now used worldwide to understand genetic variation within and between breeds, and how to effectively use it in beef cattle production (Gregory and Cundiff, 1980). Similar programs and results have also been produced and published in the Journal of Animal Science for sheep and swine through the US MARC efforts.

The 1980s was a true time of transition for livestock breeding. Computing technology had matured to the level where statistical methodology could be applied to large pedigreed performance databases—so-called BLUP methodology. Scientists successfully worked out the reduced animal model, allowing the inverse of the numerator relationship to be solved (Quaas and Pollak, 1980) and were successful in using these methods to compute for the first time what we now know as EPD and predicted transmitting ability within breeds. These new genetic evaluation tools were significantly more powerful than the previously used breeding value estimates from the sire/maternal grandsire model approach. In the time since, we have become the beneficiaries of continual refinement in genetic prediction methodology, including more accurate predictions as well as a plethora of new traits exhibiting useful levels of genetic variation (Koots et al., 1994a,b) added to the evaluation pipeline.

4

Historical Perspective—Shifting to the Era of the Genomes

The other monumental event in the 1980s was the unleashing of a new field of science called genomics. This term was first used in 1986 to collectively describe the scientific discipline of mapping, sequencing, and analysing genomic level DNA information. A technology called polymerase chain reaction (Mullis and Faloona, 1987) unleashed the forces of research into the genetic code of plants and animals. It had only taken 34 yr to go from understanding the structure of DNA to being able to start mining information from the genome!

As molecular genetics tools became available to lab scientists in the late 1980s, researchers began the arduous process of genetic mapping. Because they were unable at that time to know the base sequence of the DNA code, they had to use a somewhat "black-box" approach to identify locations on the chromosomes that might contain genes affecting these traits. This process, called linkage mapping, took advantage of DNA polymorphisms called microsatellite markers, a type of variation found readily throughout the genome. In 1994, the first genetic linkage maps of cattle detailing a few hundred markers were published by USDA-ARS scientists from US MARC (Bishop et al., 1994) and Australian Commonwealth Scientific and Industrial Research Organization (CSIRO) scientists (Barendse et al., 1994). Similar efforts have culminated in linkage maps for swine (Rohrer et al., 1994), sheep (Maddox et al., 2001), and horses (Swinburne et al., 2000). Today, these linkage maps combined with radiation hybrid and bacterial artificial chromosome maps are much better defined.

The availability of the first linkage maps allowed researchers to begin the search for regions of the genome harboring genes containing polymorphisms causing differences in performance for economically important traits—what have become known as QTL. Several livestock resource populations were formed to identify QTL affecting a wide variety of traits. The results of these projects were exciting and stimulated a considerable amount of attention in the livestock industry in the mid to late 1990s, promising that the application of marker-assisted selection was just around the corner and would soon revolutionize livestock breeding. Unfortunately, as is too often the case, the immediate promise of genomics was clearly oversold, as it has since become clear that the identification of QTL was only a first "baby step" in the process to bringing these results to a practicable technology (Dekkers, 2004).

Fortunately for the livestock genomics community, the US government had placed a high priority, through the National Institutes of Health, on better understanding the human genome, motivated by the possibility of developing new ways to combat human disease and improve human health. Initially, many of the same approaches of linkage mapping were used in human genomics, with the additional twist that model organisms, principally the laboratory mouse and rat, were studied as proxies for man, leveraging the remarkably high genomic similarity of 90% or greater between related mammalian species. Scientists also observed that although the arrangement of the pieces of the genetic puzzle was not the same across species, large regions of the genome had been conserved throughout evolution. This now allowed the opportunity to take information from species being studied with very large research budgets to infer what might be the case in livestock species using "comparative mapping." While this approach did yield results, including the identification of the callipyge condition in sheep (Cockett et al., 1994) and the myostatin gene causing double-muscling in cattle (Casas et al., 2000), only a handful of genes have been mapped to date through the "QTL-search followed by comparative mapping or fine mapping" approach.

The human genetics community quickly recognized that if progress in delivering new tools through genomics for human health applications was to occur expeditiously, building of genome sequence infrastructure was required. Linkage maps, QTL searches, comparative mapping, and some fine mapping were useful, but extremely inefficient, timely, and high in cost. Thus, in the last half of the 1990s, the National Institutes of Health, through its National Human Genome Research Institute (NHGRI), built a plan for sequencing the human genome, along with the highly used lab species of the mouse and rat. The project became broadly known as the Human Genome Project and involved a network of sequencing centres contracted to do high-throughput whole-genome sequencing. An initial rough draft of the human genome sequence was completed in 2001, followed by a complete, finished sequence in April 2003, and a detailed haplotype map in 2005, 50 yr after Watson and Crick's initial elucidation of the double-stranded helical nature of DNA. The cattle, poultry, and swine industries, however, were placed in a position to reap huge rewards from the infrastructure built by NHGRI to sequence the human genome. To build the most comprehensive infrastructure to capitalize on the human genome for discoveries in human health, NHGRI launched down a path in 2002 of supporting the sequencing of several other genomes. These were chosen to most highly leverage the investment in human genomics, as based on comparative mapping and medical model species use. Over the past 5 yr, the 6-fold or greater draft sequence assemblies have been placed in the public domain through the National Centre for Biotechnology Information, for the chicken (International Chicken Genome Consortium, 2004), bovine, and equine species. The swine genome is currently being sequenced and is expected to be completed to the same level in 2009, and plans are underway for the same to occur for the sheep genome. All of these "big science" sequencing projects have required a new form of funding and international collaboration, involving multiple governments and agencies, industry, and academia partners (Green et al., 2007).

In addition to the development of the draft sequence, each of the genome projects has also sought to develop tools that will move livestock genomics research forward quickly. In each case, additional light sequencing of animals representing additional breeds has allowed detection of large pools of new SNP. Additionally, full-length complementary DNA sequences have been developed for multiple tissue systems from the sequencing animals for the study of gene expression and further extension to expression-QTL approaches. Currently these genome sequences are being annotated through a worldwide community effort to provide a full picture of the "gene atlas" of the species.

As the genomic toolboxes were being built over the past few years, the development and release of commercial DNA tests has begun to escalate. With several companies now in the market space for this group of technologies, most with quite different business models, a variety of tests and platforms have come of age. The initial products released in the market over the past several years have generally involved small numbers of markers that have been shown to be associated with small but growing amounts of the genetic and phenotypic variation in traits of economic importance (Van Eenennaam et al., 2007).

Since 2005, the landscape in this field has been changing at a dramatic rate. This is largely because of the rapid growth in high-throughput genotyping platforms that have suddenly made it cost effective to genotype large numbers of markers in a single run. Recent efforts have culminated in the development of a panel of 56,000 bovine SNP markers (Van Tassell et al., 2008) with similar tools being developed for the chicken, swine, and sheep research communities that are expected within the next year. These developments have vastly accelerated the research and development pipeline in both public and private sectors and are poised to radically transform genomic diagnostics in the coming decade or sooner for these species. We can expect to see large panels of SNP assays made available (thousands as opposed to <100 available today) that will simultaneously provide diagnostics of multiple traits of economic interest, parentage and individual animal identification and traceability, and genetic defects of importance—all in one assay platform. Furthermore, the deployment of whole-genome selection approaches (Meuwissen et al., 2001) to livestock improvement appears to be on the threshold of reality, with recent reports for dairy cattle clearly pointing to significant enhancements in accuracy of breeding value estimates on young bulls from addition of "50K" whole-genome prediction above pedigree estimates (VanRaden et al., 200).

Collectively, the developments of the past 2 decades have brought the field of animal breeding and genetics to a new launching point. Coming to the edge of this new universe brings into clear focus the fact that the needs in the coming century are great if the field is to be able to effectively mine this new wealth of genomic information to 1) improve livestock productivity and well-being; 2) enhance natural resource stewardship while reducing the environmental footprint of production; and 3) ultimately improve the quality of human life.

Meeting Future Challenges in the Genome-Enabled ERA

Several key societal and industry shifts are currently changing the landscape of animal agriculture, including the following:

1. By 2020 to 2030, world demand for meat and dairy products is expected to increase 40 to 50% (Rosegrant et al., 2001; FAO, 2002), which will stimulate the competitiveness of food animal products.

2. An increased call from the general public for decreasing the environmental footprint of livestock and poultry production, including range, water, and air quality.

3. Competition for energy sources and feedstuffs for alternative energy production, heightening the emphasis on improved energy utilization and animal adaptability to production environments.

4. Increased attention to animal well-being and welfare, pointing out the need for robust scientific criteria to actually assess animal well-being in our production systems.

5. Increased brand differentiation, including process and historical identity of products being called for in the marketplace by retailers and their consumers.

6. Increased interest and purchasing power of consumers for products from nonconventional production systems (i.e., "natural" and "organic").

7. Recognition that we are concerned about the loss of genetic diversity, while recognizing the need to increase product uniformity and consistency. Considerable debate and disagreement in the industry regarding the value and usefulness of heterosis in the commercial sector exists for some of the livestock species.

8. The need for information continues to accelerate, with the cry for emphasis on economically relevant traits to the commercial industry, yet we have done little to put in to place evaluation for animal health, functionality, and adaptability traits (i.e., remaining heavy emphasis on output traits with not much to work with on the input side of the profit equation even though it was called for in work over 30 yr ago; Cartwright, 1970; Dickerson, 1970).

9. Structural changes in the livestock industries including consolidation, integration, and vertical coordination continue to change at a rapid rate.

10. New genomic technologies are rapidly moving the field into a brave new world of genome-enabled improvement and precision

animal management. The widespread availability of genomic technologies offers the potential to invert the genetic improvement pyramid and will likely change the delivery and structure of the breeding sector in the marketplace.

11. Mining the genome in the post-genome-sequence world is here to stay, yet we are woefully inadequate in being ready to handle all of the information that is upon us in a practically meaningful way.

The issues above, coupled with the fact that we are quickly entering the age where we have more data than we know how to effectively handle in an era where technology is outpacing our abilities to utilize it, quickly lead to the conclusion that the field of animal breeding and genetics is now entering in to a second Renaissance Age.

Although the ushering in of the genomics era has been immensely exciting scientifically, it has not come without significant and tangible costs. This research is expensive to conduct, requires large project teams, is lab intensive, and has brought to the table a host of issues associated with intellectual property rights. As research and education programs around the world geared up to make progress in this arena, other programs were often redirected to the newer arenas of molecular genetics and genomics. As a result, over the past 20 yr, the majority of traditional animal breeding positions and research herds and flocks in the public sector were eliminated or redirected into genomics and molecular biology. This has resulted in a gradual increase in the average age of animal breeding and genetics researchers and educators. Additionally, a shift has occurred in the funding for research and development with an increasingly greater percentage in this area being found in the private as compared with the public sector.

These concerns lead the USDA to recently develop a Blueprint for USDA Efforts in Agricultural Animal Genomics (USDA, 2007). This strategic plan was developed after an extensive period of information gathering and synthesis from comprehensive stakeholder groups and communities in the United States and abroad (Burfening et al., 2006; Green et al., 2007). The plan was built on the foundation that quantitative genetics has been used for many years in selecting animals for improved production (e.g., growth, yield, efficiency) and has achieved remarkable results, while pointing to the fact that addition of animal genomic technology to quantitative genetics programs has the potential to lead to more accurate and rapid animal improvement, especially for phenotypic traits that are difficult to measure (e.g., disease resistance, animal well-being, feed efficiency, product quality). Additionally, animal genomics offers new opportunities to develop precision management systems to optimize the production environment based on the genotype of the animal. The USDA Blueprint lays out a plan for how research, education, and extension efforts in animal genomics are expected to deliver the following genome-based technologies, "science to practice," to animal producers: 1) whole-genome-enabled animal selection; 2) prediction of genetic merit of individual animals from genome-based data combined with phenotypes; 3) integration of genomic data into large-scale genetic evaluation programs and the use of genomic information to design precision mating systems; 4) precision management systems to optimize animal production, health, and well-being; and 5) genomic capabilities that enable parentage and identity verification (traceability).

To deliver these outcomes, the field of traditional animal breeding and genetics will play a crucial role in addressing key priorities in the genome-enabled era. In fact, as pointed out in the USDA Blueprint, the most critical challenges in delivering value to the livestock industries from the cumulative significant investments in animal genomics are in the 2 high-need priority areas of large, highly resourced, and deeply phenotyped populations (i.e., phenomics) and in bioinformatics and computational biology. This is particularly true for the difficult-to-measure traits that have long been out of reach for genetic improvement. This includes phenotypes to genomically address the economically relevant trait areas of reproductive success and longevity in varied environments, efficiency of nutrient utilization, animal temperament, stress susceptibility, and either innate resistance or susceptibility to disease. New opportunities to alter the nutrient profile and consumer desirability of animal products also are a target in the phenomics arena. Metagenomics approaches offer the opportunity to better understand the microbial environment of the rumen and digestive environments, and, better understanding of the interaction between host and pathogen should lead to improved animal health and productivity. Opportunities for customized animal health management are now considered possible through the application of genomic profiling of animals for enhanced precision of pharmaceutical use, and pharmacogenomic approaches offer potential for development of a new generation of improved biologicals and therapeutics.

Animal breeders will need to lead the way on the integration of genomic and phenotypic data into a new era of genome-enabled animal improvement and management. Inclusion of molecular genetic information into breeding value prediction and mating system design will require new approaches, not the least of which is the ability to handle large volumes of data with efficient computing algorithms. The availability of dense marker panel information on individuals at the commercial as well as seedstock levels will require new and creative approaches to maximize the utilization of all of the potential genomic information that will be generated, from a wider sector of the gene pool. Development of customized genotyping platforms for traits within and across populations, and their continued calibration on well-phenotyped populations will be an ongoing challenge. And, it will be animal breeders working together with their colleagues in the disciplines of physiology, nutrition, muscle biology, growth and development, and immunology who will lead the way for genomic profiling to becoming a base mechanism to enable precision animal management.

As better understanding of the genomes of livestock species unfolds, animal breeders will need to determine how such current unknowns as copy number variants, rare alleles, and epigenetic effects are understood and incorporated into animal breeding programs. Genomic information will allow the field for the first time to truly estimate and understand epistatic and dominance effects and how to best utilize them in systematic programs. And, certainly it is now envisioned that full sequence of individual animals will be available long before ASAS celebrates another 50 yr of its history, driven heavily by the current emphasis of the human genetics research community on medical resequencing technologies. And last, much as predicted by leaders of the field in the past 100 yr such as Tom Cartwright, Dewey Harris, and Gordon Dickerson, it will be the new "systems biologists" who may be the legacy of quantitative genetics (Pomp et al., 2004)—who, instead of analysing the individual components or aspects of the organism, will focus on all of the components and the interactions among them, all as part of one system. They will truly seek to understand how the interactions of numerous genes, proteins, mechanisms, and the external environment of the organism produce the phenotype of an animal. While the systems biology approach will be the basic science underpinning of the phenomics revolution, the real-world application will be animal breeders working together with their colleagues in the disciplines of physiology, nutrition, muscle biology, growth and development, immunology, microbiology, and economics to lead the way for genomic profiling to become a base mechanism to enable precision animal management. Although the opportunities of this new Renaissance period of animal breeding and genetics are exciting, several daunting questions exist that must be addressed to ensure success. The reduction in number of academic animal breeding programs around the world, and particularly in North America, has resulted in a deficit of human talent and resources ready to take on these challenges. Not only are quantitative geneticists lacking, those few who are being produced in the remaining programs are entering a highly competitive job market where few are remaining in academia, with many of the trained animal breeding scientists being pulled into the plant and biomedical arenas. The need to rebuild infrastructure for developing scientists and expertise in the animal breeding area is critical and must be addressed through new strategies and models. More direct investment of private industry into the funding of these programs is beginning to occur and will need to increase in the near term. Increased reach of federal research programs into the education sector is also needed, principally through the increased availability of research funding in USDA-ARS for graduate student and postdoctoral training. The time may have come for US programs to break down past barriers that have prevented consortia-type funding so that large-scale, common sense interdependent partnerships can be instituted between federal and state agencies, universities, the private sector, and industry commodity organizations. This model has certainly had success in other areas of the world, and the area of animal breeding is one that would benefit greatly from such approaches. Finally, because of the rapid escalation in the development of genome- and phenome-based technologies, the need for public education and outreach efforts has never been greater. Again, creativity by the animal breeding community needs to be mustered to deliver this in an efficient, timely, and technologically proficient manner.

The next 100 yr of animal breeding and genetics in ASAS promises to be a truly exciting time. While the challenges are substantial, the potential and opportunities are much larger. The current and future generations of animal scientists who call themselves animal breeders will be standing on the shoulders of giants as we navigate our way forward. What we can say clearly is that we see a future that is very different from what we know today, but it will all be underpinned through diligent application and reliance on the principles of basic population and quantitative genetics coupled with the incorporation and understanding of genomic information. Opportunities abound.

The Role of Water in Animal Breeding

The role of water in animal breeding must be extended to a wider context than the animal production area, considering that 70% of the water used in the world is consumed by the whole production chain (agriculture and animal production). Therefore has a great importance the connection with other fields of the chain, as the fodder-growing and the cereal-growing, together with the evaluation and quantification of the environmental impacts. Water, that plays an essential role in the breeding, assumes different importance in relation to the animal class (birds, fish and mammals) and to the animal species. Therefore are extremely different the water requirements and the water consumptions, that are moreover strongly influenced by many factors, such as the dry matter, the climatic breeding conditions, together with the individual animal features. All that represents the starting point to determine the strategies and the ways of the water giving in animal breeding, related to the technological, project and management aspects. Besides the quantitative aspects, water must be considered as food, because it is necessary to animal survival. The importance of the quality of water used in animal breeding and its nutritional role is closely related to the qualitative characteristics and to the presence of residual and polluting substances. The animal production chain, moreover, can produce environmental impacts on the aquatic ecosystems and therefore a particular attention goes to end uses of water as output of the whole animal production chain and to the quantification of the impacts, that is extremely complicate and difficult, depending on many variables. The considerations related to animal production chain assume a different value in the productive context of the management of the water resources in the third countries.

Bayesian Estimation in Animal Breeding Using the Dirichlet Process Prior for Correlated Random Effects

In this chapter, the Dirichlet process prior was used to provide nonparametric Bayesian estimates for correlated random effects. This goal was achieved by providing a Gibbs sampler algorithm that allows these correlated random effects to have a nonparametric prior distribution. A sampling based method is illustrated. This method which is employed by transforming the genetic covariance matrix to an identity matrix so that the random effects are uncorrelated, is an extension of the theory and the results of previous researchers. Also by using Gibbs sampling and data augmentation a simulation procedure was derived for estimating the precision parameter M associated with the Dirichlet process prior. All needed conditional posterior distributions are given. To illustrate the application, data from the Elsenburg Dormer sheep stud were analysed. A total of 3325 weaning weight records from the progeny of 101 sires were used.

Breeding

Ninety-five percent of bird species are socially monogamous. These species pair for at least the length of the breeding season or—in some cases—for several years or until the death of one mate. Monogamy allows for biparental care, which is especially important for species in which females require males' assistance for successful brood-rearing. Among many socially monogamous species, extra-pair copulation (infidelity) is common.

Such behaviour typically occurs between dominant males and females paired with subordinate males, but may also be the result of forced copulation in ducks and other anatids. For females, possible benefits of extra-pair copulation include getting better genes for her offspring and insuring against the possibility of infertility in her mate. Males of species that engage in extra-pair copulations will closely guard their mates to ensure the parentage of the offspring that they raise.

Other mating systems, including polygyny, polyandry, polygamy, polygynandry, and promiscuity, also occur. Polygamous breeding systems arise when females are able to raise broods without the help of males. Some species may use more than one system depending on the circumstances.

Breeding usually involves some form of courtship display, typically performed by the male. Most displays are rather simple and involve some type of song. Some displays, however, are quite elaborate. Depending on the species, these may include wing or tail drumming, dancing, aerial flights, or communal lekking. Females are generally the ones that drive partner selection, although in the polyandrous phalaropes, this is reversed: plainer males choose brightly coloured females. Courtship feeding, billing and allopreening are commonly performed between partners, generally after the birds have paired and mated.

Homosexual behaviour has been observed in males or females in numerous species of birds, including copulation, pair-bonding, and joint parenting of chicks.

Breeding in the Wild

Breeding in the wild is the natural process of animal reproduction occurring in the natural habitat of a given species. This terminology is distinct from animal husbandry or breeding of species in captivity. Breeding locations are often chosen for very specific requirements of shelter and proximity to food; moreover, the breeding season is a particular time window that has evolved for each species to suit species anatomical, climatic and other ecological temporal events. Many species migrate considerable distances to reach the requisite breeding locations. Certain common characteristics apply to various taxa within the animal kingdom, which traits are often sorted among amphibians, reptiles, mammals, avafauna, arthropods and lower life forms. For many amphibians an annual breeding cycle applies, typically regulated by ambient temperature, precipitation, availability of surface water and food supply. This breeding season is accentuated in temperate regions, where prolonged aestivation or hibernation renders many amphibian species inactive for prolonged periods. Breeding habitats are typically ponds and streams.

Camargue Cattle

The Camargue cattle breed, in Provençal: Raço di biòu, is native to the Camargue marshlands of the delta of the River Rhone in southern France. The cattle are black in colour with upward sweeping horns. They are hardy animals thriving on the marshes where they live semi-wild, tended by the mounted herders called gardians who ride the famous Camargue horses which live in the same area. The bulls are used for bull-fighting and for the *course camarguaise*; animals suitable neither for the bull-ring nor for breeding are sold for beef. The breed is also regarded as a tourist attraction. The meat of the Camargue breed, along with that of the Brava cattle breed and crosses between the two, can under strict conditions of pasturage and of zone and methods of production be marketed with the Appellation d'origine contrôlée (AOC) "Taureau de Camargue" certification of origin; animals that have appeared in the bull-ring are excluded. The breed is not endangered, with the population estimated at 5950 head. The cattle share the wetland environment with the horses, wild boar, and flamingoes. Cattle Egrets live with the cattle.

Belgian Hare

The Belgian Hare is a fancy breed of domestic rabbit, that was developed through selective breeding to closely resemble the wild hare in physical appearance, and is believed to be one of the most intelligent and active breeds of domestic rabbit. Averaging 6 to 9 pounds in weight, the Belgian Hare is characterized by its long, slender body and agile legs that closely resemble a hare, and can live up to ten years or more.

The first Belgian Hares were bred in Belgium in the early 18th century out of selective breeding between domestic and wild European rabbits, with the intent of creating a practical meat rabbit. In 1874, they were imported to England and called the "Belgian Hare." English breeders made the Belgian Hare appear more spirited, like wild English rabbits. By 1877 the first Belgian Hares were shown in the America, where it immediately rose in popularity, giving rise to thousands of Belgian Hare clubs around the country, thousands were bred, and some sold for as much as 1,000 US dollars.

The first of these clubs was known as the "American Belgian Hare Association". With a wide and scattered membership the club lasted not much more than a year. In 1897 the "National Belgian Hare Club" was formed. Twelve years after the formation of the National Belgian Hare Club of America, and as additional breeds were introduced in the US, a new "all-breed" club, the "National Pet Stock Association" was formed. After several name changes, the National Pet Stock Association became the American Rabbit Breeders Association As years passed, the National Belgian Hare club of America also passed from existence. In June, 1972, a group of Belgian Hare breeders gathered together to apply for a specialty club charter from the American Rabbit Breeders Association to replace the National Belgian Hare Club of America. In July, 1972, the charter was granted and the last, and most prominent of these groups, the "American Belgian Hare Club" was established, that continues to exist to this day.

In 1917, their popularity began to fade away, and one of the reasons attributed to this decline is the failed attempt by many breeders to turn the Belgian Hare, a naturally race rabbit, into a meat rabbit, a role to which they were physically and behaviourally unsuited. However, today, true Belgian Hares are rare, due partly to the degree of difficulty many have had in breeding them

Breeding

The ideal age for the female Belgian Hare to start breeding is about 9 months of age, and can produce large litters of between 4 to

8 babies, with a gestation period of between 28 to 31 days. On average, they give birth at 30 to 32 days.

As different bucks and does react to the breeding situation differently, one must be prepared to adapt their breeding practices to the Hare's preferences.

Like all rabbits, Belgian Hares are induced ovulators (a doe can immediately ovulate at the time of breeding). However, the doe has a 10-14 day receptivity cycle. During her receptivity period she will accept the buck readily. A receptive doe can be identified by the dark pink, moist, appearance of her vulva. If you have a receptive doe to breed, by all means, try placing her in your buck's cage first, but watch them carefully. In many instances, the buck, the doe, or both animals can become so upset by the appearance of the other Hare, that they will either attack (and you will have a real fight on your hands) or retreat to a corner and cower.

A method of breeding used very successfully by some breeders is the "honeymoon cottage." In this method, a large cage, at least 60 inches long is partitioned into two parts with a plywood wall. The smaller part should be about 18 inches long, and the larger, 42 inches. The plywood wall should have a 6 inch round hole through it at the doe's shoulder height. A clean, sterilized cage should be used so that there are no other animal odors on it. In other words, the cage should be "neutral ground." It is recommended that the doe and buck be placed together for 10 – 14 days to bond with each other, and eventually mate. Initially, it is expected that may chase each other around back and forth through the hole, and after a while, the doe will discover that when the buck's amorous advances became too much, she can defend her territory, i.e., her side of the cage, by standing with her head in the hole. They will mate when both buck and does are ready. Ten to fourteen days through her gestation period, the buck should be removed, and the larger part of the cage should be cleaned, leaving the cardboard covered smaller part alone. After a while, the doe will make a nest on the floor of the darkened, smaller, part of the cage, and conceive at the end of her gestation period. Following a period of eight weeks after the conception, it is recommended that the litter be weaned by removing the doe, and the doe and buck be separated.

The domestic chicken is descended primarily from the Red Junglefowl (Gallus gallus) and is scientifically classified as the same species. As such it can and does freely interbreed with populations of red jungle fowl. Recent genetic analysis has revealed that at least the gene for yellow skin was incorporated into domestic birds through hybridization with the Grey Junglefowl (G. sonneratii). The traditional poultry farming view is stated in Encyclopaedia Britannica (2007): "Humans first domesticated chickens of Indian origin for the purpose of cockfighting in Asia, Africa, and Europe. Very little formal attention was given to egg or meat production... " In the last decade there have been a number of genetic studies. According to one study, a single domestication event occurring in the region of modern Thailand created the modern chicken with minor transitions separating the modern breeds. However, that study was later found to be based on incomplete data, and recent studies point to multiple maternal origins, with the clade found in the Americas, Europe, Middle East, and Africa, originating from the Indian subcontinent, where a large number of unique haplotypes occur.

It has been claimed (based on paleoclimatic assumptions) that chickens were domesticated in Southern China in 6000 BC. However, according to a recent study, "it is not known whether these birds made much contribution to the modern domestic fowl. Chickens from the Harappan culture of the Indus Valley (2500-2100 BC), in what today is Pakistan, may have been the main source of diffusion throughout the world."

A northern road spread chicken to the Tarim basin of central Asia. The chicken reached Europe (Romania, Turkey, Greece, Ukraine) about 3000 BC. Introduction into Western Europe came far later, about the 1st millennium BC. Phoenicians spread chickens along the Mediterranean coasts, to Iberia. Breeding increased under the Roman Empire, and was reduced in the Middle Ages. Middle East traces of chicken go back to a little earlier than 2000 BC, in Syria; chicken went southward only in the 1st millennium BC. The chicken reached Egypt for purposes of cock fighting about 1400 BC, and became widely bred only in Ptolemaic Egypt (about 300 BC). Little is known about the chicken's introduction into Africa. Three possible ways of introduction in about the early first millennium AD could have been through the Egyptian Nile Valley, the East Africa Roman-Greek or Indian trade, or from Carthage and the Berbers, across the Sahara. The earliest known remains are from Mali, Nubia, East Coast, and South Africa and date back to the middle of the first millennium AD. Domestic chicken in the Americas before Western conquest is still an ongoing discussion, but blue-egged chicken, found only in the Americas and Asia, suggest an Asian origin for early American chickens.

A lack of data from Thailand, Russia, the Indian subcontinent, Southeast Asia and Sub-Saharan Africa makes it difficult to lay out a clear map of the spread of chickens in these areas; better description and genetic analysis of local breeds threatened by extinction may also help with research into this area.

Current

Under natural conditions, most birds lay only until a clutch is complete, and they will then incubate all the eggs. Many domestic hens will also do this–and are then said to "go broody". The broody hen will stop laying and instead will focus on the incubation of the eggs (a full clutch is usually about 12 eggs). She will "sit" or "set" on the nest, protesting or pecking in defense if disturbed or removed, and she will rarely leave the nest to eat, drink, or dust-bathe. While brooding, the hen maintains the nest at a constant temperature and humidity, as well as turning the eggs regularly during the first part of the incubation. To stimulate broodiness, an owner may place many artificial eggs in the nest, or to stop it they may place the hen in an elevated cage with an open wire floor.

At the end of the incubation period (about 21 days), the eggs, if fertile, will hatch. Development of the egg starts only when incubation begins, so they all hatch within a day or two of each other, despite perhaps being laid over a period of two weeks or so. Before hatching, the hen can hear the chicks peeping inside the eggs, and will gently cluck to stimulate them to break out of their shells. The chick begins by "pipping"; pecking a breathing hole with its egg tooth towards the blunt end of the egg, usually on the upper side. The chick will then rest for some hours, absorbing the remaining egg yolk and withdrawing the blood supply from the membrane beneath the shell (used earlier for breathing through the shell). It then enlarges the hole, gradually turning round as it goes, and eventually severing the blunt end of the shell completely to make a lid. It crawls out of the remaining shell, and its wet down dries out in the warmth of the nest.

A day-Old Chick

The hen will usually stay on the nest for about two days after the first egg hatches, and during this time the newly hatched chicks live off the egg yolk they absorb just before hatching. Any eggs not fertilized by a rooster will not hatch, and the hen eventually loses interest in these and leaves the nest. After hatching, the hen fiercely guards the chicks, and will brood them when necessary to keep them warm, at first often returning to the nest at night. She leads them to food and water; she will call them to edible items, but seldom feeds them directly. She continues to care for them until they are several weeks old, when she will gradually lose interest and eventually start to lay again.

Modern egg-laying breeds rarely go broody, and those that do often stop part-way through the incubation. However, some "utility" (general purpose) breeds, such as the Cochin, Cornish and Silkie, do regularly go broody, and they make excellent mothers, not only for chicken eggs but also for those of other species—even those with much smaller or larger eggs and different incubation periods, such as quail, pheasants, turkeys or geese. Chicken eggs can also be hatched under a broody duck, with varied success.

Bear Breeding

The age at which bears reach sexual maturity is highly variable, both between and within species. Sexual maturity is dependent on body condition, which is in turn dependent upon the food supply available to the growing individual. In the females of smaller species may have young in as little as two years, whereas the larger species may not rear young until they are four or even nine years old. First breeding may be even later in males, where competition for mates may leave younger males without access to females.

The bear's courtship period is very brief. Bears in northern climates reproduce seasonally, usually after a period of inactivity similar to hibernation, although tropical species breed all year round. Cubs are born toothless, blind, and bald. The cubs of brown bears, usually born in litters of 1–3, will typically stay with the mother for two full seasons. They feed on their mother's milk through the duration of their relationship with their mother, although as the cubs continue to grow, nursing becomes less frequent and cubs learn to begin hunting with the mother. They will remain with the mother for approximately three years, until she enters the next cycle of estrus and drives the cubs off. Bears will reach sexual maturity in five to seven years. Male bears, especially Polar and Brown Bears, will kill and sometimes devour cubs born to another father in order to induce a female to breed again. Female bears are often successful in driving off males in protection of their cubs, despite being rather smaller.

Winter Dormancy

Many bears of northern regions are assumed to hibernate in the winter. While many bear species do go into a physiological state often colloquially called "hibernation" or "winter sleep", it is not true hibernation. In true hibernators, body temperatures drop to near ambient and heart rate slows drastically, but the animals periodically rouse themselves to urinate or defecate and to eat from stored food. The body temperature of bears, on the other hand, drops only a few degrees from normal and heart rate slows only slightly. They normally do not wake during this "hibernation", and therefore do not eat, drink, urinate or defecate the entire period. Higher body heat and being easily roused may be adaptations, because females give birth to their cubs during this winter sleep

Method for Breeding Animals

A method for breeding an animal, wherein said method comprises:administering a feed composition (a) to the animal in an initial breeding stage; andadministering a feed composition (b) to the animal after completion of the initial breeding stage,wherein the feed composition (a) comprises a silver-carrying zeolite having a silver content of 0.1-5.0 wt. %, based on a total weight of the silver-carrying zeolite, wherein the feed composition (a) comprises silver in an amount of A wt. %, based on a total weight of the feed composition (a), wherein A is 0.000001-0.2 wt. %,wherein the feed composition (b) comprises a silver-carrying zeolite having a silver content of 0.1-5.0 wt. %, based on a total weight of the silver-carrying zeolite, wherein the feed composition (b) comprises silver in an amount of B wt. %, based on a total weight of the feed composition (b), wherein B is less than A, and where in the animal is selected from the group consisting of fowl bred for egg collection and dairy cattle.

A method for breeding an animal, wherein said method comprises: administering a feed composition (a) to the animal in an initial breeding stage; administering a feed composition (b) to the animal during a term extending from after completion of the initial breeding stage to 5-20 days before a distribution or shipping date of the animal; and administering a feed composition (c) to the animal during a term between 5-20 days before the distribution or shipping date and the distribution or shipping date of the animal, wherein the feed composition (a) comprises a silver-carrying zeolite having a silver content of 0.1-5.0 wt. %, based on a total weight of the silver-carrying zeolite, wherein the feed composition (a) comprises silver in an amount of A wt. %, based on a total weight of the feed composition (a), wherein A is 0.000001-0.2 wt. %,wherein the feed composition (b) comprises a silver-carrying zeolite having a silver content of 0.1-5.0 wt. %, based on a total weight of the silver-carrying zeolite, wherein the feed composition (b) comprises silver in an amount of B wt. %, based on a total weight of the feed composition (b), wherein B is less than A, wherein the feed composition (c) comprises a silver-carrying zeolite having a silver content of 0.1-5.0 wt. %, based on a total weight of the silver-carrying zeolite, wherein the feed composition (c) comprises silver in an amount of C wt. %, based on a total weight of the feed composition (c), wherein C is greater than B, and where in the animal is one for obtaining meat.

The method according to claim 2, wherein the animal is selected from the group consisting of poultry, cattle, swine, wild boar, equine, sheep and deer.

The method according to claim 2, wherein the animal is selected from the group consisting of domestic fowl bred for obtaining meat, beef cattle and swine.

Genomics and Beyond

An organism's Genome is its Genetic Material (DNA)

Animal genomics, broadly defined, is the study of animal DNA, its organisation into genes, the roles and interactions of gene products (mostly proteins), the control of gene expression, and ultimately all downstream impacts on animals, their traits, and their interactions with the environment. The sequencing of the human genome was a milestone in the development of modern biotechnology, opening new avenues for understanding the control of complex traits at a molecular level. The genomes of several farm animal species (cattle, pig, rabbit, and aquacultural species) are also being sequenced and analysed. This will facilitate the integrated analysis of biological functions.

At sequence level, genomic research is yielding genetic markers for selection, pedigree control, and traceability. It is facilitating the identification of trait genes for use in breeding programmes. Beyond this level opens the whole new realm of "omics" research: transcriptomics (focus: expressed genes), proteomics (proteins and their functions and interactions), and metabolomics (metabolites and metabolic pathways).

This research is bridging (part of) the gap between 'structural' genomic knowledge (markers, maps, sequences, genes) and everything we hope to learn by exploiting it. It provides a bridge to nutrition, health, and more. It is yielding precious knowledge on a wide range of processes related to health and food safety, such as acquired resistance to contamination through gut health or the long-term effects of an altered maternal diet (foetal programming) in both animals and humans. The application of such knowledge (in marker-assisted breeding programmes, for instance) will contribute to improving animal health and food safety, animal welfare, and the biodiversity of breeding populations. In the long run it will boost the competitiveness of European farm animal breeding.

Functional Genomics

This area includes transcriptomics and proteomics. It aims to elucidate gene functions. Transcriptomics uses powerful tools (e.g. microarrays: arrays of material resulting from gene transcription, the first step in gene expression) to study simultaneously the expression of many (ultimately all) genes in a genome in various situations (specific developmental stages, environmental stress...). Proteomics exploits various technologies to study which proteins are present in ? or secreted by ??different cells under different conditions, and also looks at protein interactions.

Functional genomics is used to study genes that contribute jointly to a specific function as part of the 'interactome' (the whole set of molecular interactions in cells). This notably includes networks of interacting proteins, structural and mitochondrial DNA, and inhibitor RNAs (RNA is quite similar chemically to DNA; it plays multiple roles, notably in gene expression and its regulation). By combining multiple approaches and tools, it should eventually be possible to build models linking together large numbers of genes and their products, in order to understand the impact of gene variation on animal traits, mediated through the relevant metabolic pathways. To achieve this, a considerable research effort is needed, yielding better understanding of animal biology on the basis of knowledge obtained from the genomes of model organisms and humans.

Exploring Within-Species Variability

To allow effective selection of animals via genomic approaches, it is necessary to explore within-species variability and to quantify factors such as linkage disequilibrium (the tendency of genetic features that are close to each other on the genome to be inherited together, and thus frequently to be found together in the animals composing a population). Effective genomic selection requires the development of tools for genotyping animals by marker typing or direct sequencing. It requires improved tools for phenotyping animals more exhaustively and quickly so as to identify genes that control variation of multiple traits. We also need to develop strategies for predicting phenotypes on the basis of genotypes. Here research should notably focus on the genetic control of variation in gene expression and the identification of genome regions undergoing epigenetic modifications such as imprinting. Given the strong negative impact of inbreeding on reproductive and health traits, we need to find ways to reduce the inbreeding coefficient in pigs, poultry, aquaculture species, and dairy cattle.

Exploring Between-Species Variation

By learning more about between-species variation, it will become clear which vital processes program which functions in animals. This will enable breeders to identify animals yielding healthy animal products without this being detrimental to their own health.

With genome-scale sequencing of domestic species, between-species comparisons (including comparisons with mouse and human sequences) will make it possible to identify variability 'hot spots' in genes and/ or genome regions so as to focus the search for single nucleotide polymorphisms (SNPs) responsible for substitutions in protein sequences in domestic animals (SNPs are sites where a single 'letter' in the 'text' of a gene is variable).

Numerical Biology

Future biology will increasingly require the analysis of large volumes of data (e.g. genomic data). Better numerical methods for analysis and modelling are needed to address a whole range of biological problems from the molecular to the ecosystem level. Identifying functional variants in DNA sequences or differentially expressed genes on microarrays, analysing host-pathogen interactions in farm animals populations, are just a few examples among many.

There will be an increasing need for better numerical models of biological (sub-)systems. For example, scientists currently use separate models to predict protein folding from amino acid sequences and the spread of disease from epidemiological data. A more elaborate model would predict the impact of designed changes in an immune response gene on the frequency of disease epidemics in domestic animals.

As the need for numerical analysis and modeling spans all areas of biology, the field of animal breeding and genetics will have to adapt to its own needs some new methods developed for other purposes, in addition to contributing approaches specifically designed for livestock problems.

Biology of Systems and Traits

Currently the basic biology (genetics, biochemistry, and physiology) underlying the variability of most traits of interest to animal breeders is poorly understood. This is an obstacle to controlling physiological processes that influence such traits.

Molecular genetics will yield abundant novel information on the fundamental regulation of such processes. Here research must aim to provide detailed operational understanding of the pathways involved.

Many gene products influence several metabolic pathways, and most metabolic pathways are influenced by several gene products. This makes for very complicated physiological network patterns. In addition, whole sets of genes can be switched on or off via so-called 'epigenetic' mechanisms (DNA methylation, histone acetylation, RNA interference) participating in complex regulatory networks. Such mechanisms are increasingly found to play significant roles in phenotype development and transmission.

Control of a single gene may thus cause unwanted side effects when the physiological network around it is not sufficiently understood. Future research must focus on dissecting the genetic basis of traits of relevance to sustainable animal breeding and the Inter-relationships of these traits.

Whole-Animal Biology and Gene-by-Environment Interactions (GxE)

Understanding individual traits is not enough. What matters is the biology of the whole animal and how an animal's genes interact with its living environment to determine its quality of life and performance.

Biological research is already moving towards predictive biology. A first step is to model the various systems that constitute the biology of a cell (creating an e-cell) so as to predict cell responses to environmental change (e.g. altered nutrient supply).

The next step is to build models that describe cell-cell interactions in tissues and finally to build upward from tissues to a whole animal. This is a considerable challenge, but it is undeniably the direction in which biology is heading. High-throughput laboratory methods now make it possible to measure many components of a system in parallel, and these approaches complement the traditional whole-animal research supporting farm animal production. With modern research techniques, scientists will be able to revisit 100 years of animal science with a view to expanding knowledge of all biological mechanisms.

Future research must focus on developing an integrated understanding of the lifetime functioning of whole animals and their interactions with the environment. At each step it will be essential to evaluate progress in the light of breeding reality. What breeders need especially is a detailed understanding of relationships between performance, health, welfare, and environmental impact combined with costeffective, robust, reliable ways to measure traits reflecting these relationships in animal populations.

Population Biology

Future qualitative and quantitative genetic technology will give breeders much control over the genetic makeup of individual animals selected for breeding. In other words, what can be selected with increasing precision is the genome of an individual. Yet population-level effects also deserve consideration for at least three reasons. Firstly, social interactions in groups of animals are associated with behavioural repertoires that are important for animal welfare and for proper functioning of the group: maternal behaviour, dominance and aggression... To the extent that such interactions are linked to production traits, selection could have negative side effects when the physiology of social behaviour is not sufficiently understood.

Secondly, the production potential inherent in an animal's genes is variably expressed according to the nutritional, climatic, infectious, and social environment. This is called environmental sensitivity or phenotypic plasticity. It provides a measure of genotype-environment interactions at the level of an individual animal. Future genetic technology will offer increasing control over environmental sensitivity, and again side effects are likely if interaction mechanisms are not properly understood at population level.

It is important to take into account the dynamics of the environment in which future breeding animals will have to produce. The environment will change in the next twenty years and strategies for the genetic improvement of animal populations must adapt to these changes. Also important is the development of areas such as the improved management of pastures and silvo-pastoral systems for increased overall efficiency in animal production. Research should focus on defining, through population biology, which will be the most efficient genotypes for predictable environments.

Thirdly, disease transmission is perhaps the most crucial population-level issue. Infectious disease involves two (or more) 'actors': host (the farm animal) and pathogen. These two players interact through infection (pathogen to host) and immunity (host to pathogen), and the pathways involved in both processes are under genetic control. Future genetic technology will provide much knowledge about regulation on each side of such a system, but the connection between the two, i.e. the host-pathogen interaction, is likely to be the crucial element in getting the system under control. Exploring this connection means focusing on epidemiology in relation to pathogen virulence and host resistance. It requires studying the co-evolution of hosts and pathogens.

Research exploiting emerging technologies is bound to provide much knowledge on the regulation of the immune system at the individual animal level, but turning this knowledge into operational control will require linking it properly to population-level epidemiology.

Future research must focus on understanding how variation in a population affects the overall performance of groups of animals and how changes in biodiversity are likely to influence the control of infectious disease. This will ultimately lead to defining appropriate levels of biodiversity for the long-term sustainability of animal agriculture and aquaculture.

The Effects of Genetic Groups

Brazil has the largest commercial cattle herd in the world with approximately 170 million animals and a production of approximately 8.5 million tons of carcass each year (Anualpec, 2008). From this total, about 30% (2.5 million tons) is exported to several countries around the world. The consumer market for beef has become increasingly demanding as a result of negative factors associated with meat production and quality (Saucier, 1999). Among these factors is the relationship between beef consumption and heart disease, atherosclerosis, intestinal cancer and obesity, among other diseases.

There are several factors that influence the carcass characteristics, chemical composition and fatty acid profile in cattle. Genetic groups, nutrition, finishing systems and gender (Padre et al., 2007) are some reasons for the variations in meat quality. Genetic group is one of the most important factors for fat deposition and composition, which needs to be understood because of its genetic transmission. However, the detailed mechanisms of this variation, and whether or how they can be manipulated are not clearly known. The aim here is to review some characteristics that can change the meat quality (carcass characteristics, chemical composition and fatty acid profile).

Carcass Characteristics

There are several factors that are responsible for the carcass quality. Carcass conformation, carcass length, leg length, cushion thickness, Longissimus muscle area, fat thickness, colour, texture and marbling are characteristics responsible for carcass quality. The most important factor responsible for change in the carcass characteristics is the genetic group. Some papers have shown that different genetic groups influence the carcass characteristics.

Chemical Composition

The chemical composition is represented by several factors such as: moisture, ash, crude protein, total lipids and total cholesterol.

Variations in moisture percentage occur when there is a variation in lipid percentage in Longissimus muscle.

Ash percentage is little influenced by nutrition or genetic group. Ash is important for the supply of sodium, potassium, phosphorus and magnesium which are of great nutritional importance to humans.

Some authors reported crude protein percentage in Longissimus muscle varying between 21 and 24%. Thus, genetic groups, nutrition and gender can alter crude protein percentage in Longissimus muscle of bovines.

Total lipids in Longissimus muscle of cattle can vary from 2 to 5%. This is the parameter that is most influenced by genetic groups, nutrition and gender.

Cholesterol is an essential compound for the body; it takes part in the synthesis of hormones and bile salts, with half of total cholesterol produced endogenously and the remainder derived from the diet (Campbell, 1995). Until 50 years ago, the solution for reducing blood cholesterol levels seemed simple: the recommendation was to eat foods low in cholesterol, such as replacing butter with vegetable margarine and eating less meat and eggs. However, after advances in the knowledge of the functions of the human body, it is now known that the amount of cholesterol in food does not necessarily determine blood cholesterol levels. The liver synthesizes and stores cholesterol and these processes are regulated by body needs and cholesterol availability in the diet (Campbell, 1995).

Fatty Acid Profile

Saturated, monounsaturated and polyunsaturated fatty acids.

Fatty acids are "amphiphilic", i.e. they have a carboxyl group (hydrophilic) at the polar end and a hydrocarbon chain at the non-polar tail (hydrophobic) (Webb et al., 2008). Saturated fatty acids consist of only single bonds (Campbell, 1995) and unsaturated fatty acids have double bonds between carbons. Polyunsaturated fatty acids have more than one double bond. In vegetable oils there is a predominance of polyunsaturated fatty acids. On the other hand, in animal fat, there is a predominance of saturated fatty acids.

The softness of the lipids is determined by two factors: the length of the predominant fatty acid chain and the presence or absence of double bonds (Campbell, 1995). In animal fat, there is a predominance of long chain fatty acids, with twelve or more carbons atoms.

Essential Fatty Acids in Diet

Some fatty acids are required in the human diet because they cannot be synthesized by the human organism (Webb et al., 2008) These fatty acids function as carriers of the fat soluble vitamins (Vitamin A, D, E and K) and realize an important role in the immune response of the animal organism (Webb et al., 2008). These fatty acids can be found in meat of cattle. Essential fatty acids are: linolenic acid (C18:3 n-3), linoleic acid (C18:2 n-6), arachadonic acid (C20:4 n-6), eicosapentanoic acid (C20:5 n-3) and docosahexaenoic acid (C22:6 n-3).

Non-Essential Fatty Acids in Diet

The non-essential fatty acids are those fatty acids that can be synthesized by the animal organism (Webb et al., 2008). The major non-essential fatty acids found are saturated fatty acids (C16:0 and C18:0) and the monounsaturated fatty acid C18:1 n-9 (Prado et al., 2008a; b; c; d).

Conjugated Fatty Acid (CLA)

Trans unsaturated fatty acid is an important precursor for the formation of conjugated linoleic acid (CLA) in tissues (Webb et al., 2008) Due to its properties as an intermediary in the biohydrogenation process of linoleic acid (C18:2 n-6) in the rumen, this fatty acid can be transformed into CLA (C18:2 c-9, t-11) in the tissues of ruminants by the delta-9-desaturase enzyme after being absorbed (Griinari et al., 2000).

Genetic Groups

Genetic group is an important factor that affects the meat quality (carcass characteristics, chemical composition and fatty acid profile) in cattle. For example, British cattle are well known for their highly marbled meat, while the zebu breed contains less fat and more connective tissue.

Genetic variability consists of differences among species, breeds or lines, differences due to the crossing of breeds, and differences among animals within breeds (Perotto et al., 2000). The latter source of variation is estimated by heritability and genetic correlations. Breed effects may be influenced by the segregation of major genes, one of which is the double-muscled gene in cattle. It is sometimes difficult to assess the real contribution of genetics to differences in meat quality (Webb, 2006). Breed comparisons are often confounded by other effects, such as fat level, live weight, age at slaughter and production system (Webb, 2006).

Carcass Characteristics

The animals used to compare carcass characteristics were finished in a feedlot and they had a similar slaughter age (20 to 22 months). In this review, animals were used from the follow genetic groups: NellorexAberdeen Angus (NelxAng), CanchinxAberdeen Angus (CanxAng), CharolaisxCaracu (ChaxCar), Puruna first generation (Pur1), NellorexContinental breed (NelxCon), Puruna second generation (Pur2), PurunaxBritish breed (PurxBri) and PurunaxCanchin (PurxCan).

Final weight (FWE) ranges between 450 to 558 kg. NelxAng presents the higher FWE. This is the result of heterosis and of differences in the original breeds used for crossbreeding (Prado et al., 2008a). In this respect, the presence of genes from Aberdeen Angus in general, increased the advantage of these cattle. In the same way, hot carcass weight (HCW) ranges between 229 to 304 kg and NelxAng presents the higher value for HCW.

Conformation (CON), fat thickness (FAT), texture (TEX) and marbling (MAR) are also influenced by genetic groups. These characteristics account for more than 35% of variation. On the other hand, hot carcass dressing (HCD), carcass length (CAL), leg length (LEL), cushion thickness (CUT), Longissimus muscle area (LMA) and colour (COL) account for less than 20% of variation due to genetic groups.

Genetic groups that have zebu genes in their composition show higher HCD than genetic groups without zebu breeds. Prado et al. (2008b) suggest that heterosis is responsible for the higher value of HCD. However, CON is lower when zebu genes compound the genetic group. Prado et al. (2008b) affirm that genetic groups with zebu genes present a lower conformation because zebu breeds show irregular conformation.

FAT is a characteristic that represents excess energy stored as subcutaneous fat. FAT is influenced by genetic groups and its range is between 2.70 to 5.10 mm.

MAR is an important characteristic to meat export because some countries, such as the United States of America, Japan and China, prefer meat with much marbling. This parameter is influence greatly by genetic group. In this review, MAR ranges between 4.13 to 7.78 points. The score for this parameter is 1 to 18 points (Muller, 1980). Genetic groups that have European breeds in their composition have higher MAR scores. On the other hand, genetic groups that have zebu breeds in their composition show lower MAR scores.

Chemical Composition

Moisture percentage has little variation due to genetic groups, and changes from 72.8 to 74.2%. Prado et al. (2008b) suggest that variations in moisture percentage occur when there is variation in percentage of total lipids. However, in this review, percentage of total lipids changes 170% among genetic groups.

Ash percentage has a variation around 12% due to genetic groups. Ash percentage ranges between 1.01 to 1.13%.

In the same way, crude protein percentage changes around 9%. In this review, the lowest value is 22.8% and the highest value is 24.9%. Prado et al. (2008c) suggest that crude protein percentage is not altered by genetic groups.

Total lipids percentage presents variation due to genetic groups. In general, total lipid levels in the Longissimus muscle of cattle finished in a feedlot is close to 3% (Rotta et al., 2009). Animals from genetic groups CanxAng (3.91%) present the higher value for this parameter; on the other hand, animals from genetic groups ChaxCar (1.45%) present the lower value. However, percentage of total lipids observed in genetic groups is below the maximum level ([+ or-] 5%) regarded as acceptable for the prevention of diseases related to fat content in beef, according to recommendations from the English Health Department (HMSO, 1994).

Total cholesterol presents a variation of around 55% due to genetic groups. It changes between 34.1 to 52.9 mg/100 g in Longissimus (Prado et al., 2008a; b; c). Prado et al. (2009b) suggest that the presence of Zebu genes in the makeup of genetic groups appears to result in higher total cholesterol levels in Longissimus muscle, perhaps as a result of an increase in muscle membranes. Nevertheless, the average total cholesterol level observed is near that regarded as harmful to human health, which is set to [greater than or equal to] 50 mg/100 g of muscle (Pensel, 1998; Saucier, 1999).

Fatty Acid Profile

Genetic groups influence variation of fatty acid profile. Variation of around 60% can be observed in 14:0 (1.57 to 2.73%), 16:1 n-7 (1.72 to 2.94%), 17:0 (0.59 to 0.97%), 18:1 t-11 (0.85 to 2.50%), 18:2 n-6 (2.26 to 7.25%), 18:3 n-6 (0.01 to 0.20%), 18:3 n-3 (0.12 to 0.73%), 18:2 c-9 t-11 (0.17 to 0.49%), 20:4 n-6 (0.59 to 2.29%), 20:5 n-3 (0.09 to 0.68%), 22:5 n-3 (0.07 to 0.81%) and 22:6 n-3 (0.06 to 0.46%). Prado et al. (2008b; c) suggest that the major factor associated with the variation in fatty acid profile is the genetic groups.

The fatty acids 12:0, 14:0 and 16:0 are considered hypercholesterolemic; they are responsible for the increase in quantity of lipoproteins of low density—LDL (low density lipoprotein)—that are responsible for heart disease (Souza et al., 2006). So, foods with low amounts of these fatty acids are required. In Longissimus muscle of cattle the percentage of 14:0 and 16:0 is considered high (35% of the total fatty acids). The genetic group NelxCon presents the lowest levels to these fatty acids.

Though the fatty acid 18:2 n-6 is considered an essential fatty acid, it is related to the imbalance of n-6/n-3 ratio. This occurs by high presence of this fatty acid in the Longissimus muscle of cattle in relation to n-3 fatty acids (Rotta et al., 2009). The highest percentage of this fatty acid is found in the genetic group NelxAng (7.25%) and the lowest percentage is present in the genetic group Pur2 (2.26%).

The fatty acid 18:3 n-3 is considered strictly essential and the presence of this fatty acid is important due to the capacity to form other important fatty acids (Wood et al., 2003). This fatty acid is found in the highest percentage in the genetic group ChaxCar (0.73%).

Saturated fatty acid (SFA) and monounsaturated fatty acid (MUFA) show variation around 15% among different genetic groups. On the other hand, polyunsaturated fatty acids (PUFA), n-6, n-3, PUFA/SFA and n-6/n-3 present variation due to different genetic groups.

PUFA has a variation from 3.99 to 10.9%. Genetic groups that have zebu genes in their composition show high percentages of PUFA. Prado et al. (2008b; c) suggest that zebu genes are responsible for more deposition of connective tissue which has a high content of polyunsaturated fatty acids. n-6 percentage ranges between 3.44 to 9.23%. The genetic groups that have the zebu breed in their composition show high percentages of this category of fatty acids. This is due to these genetic groups having more PUFA in their fatty acid profile.

Variation is present for n-3 percentage due to different genetic groups. The lower value found for n-3 is 0.48% and the higher value is 2.56%. The variation in fatty acid profile is responsible for variation of n-3 percentage.

PUFA/SFA ranges between 0.08 to 0.22. PUFA/SFA plays an important role in reducing the risk of coronary heart disease (Hu, 2001). The Department of Health (1994) recommends a ratio of. 0.45. The genetic groups that are near this value are those with zebu genes in their composition. n-6/n-3 ranges between 2.70 to 11.1. Only two genetic groups (ChaxCar (2.70) and PurxBri (3.77)) show ideal ratios. This led to the n-6/n-3 ratio in the Longissimus muscle being considered high when compared to the maximum of 4.0 recommended by the Department of Health (1994).

Nutrition

Nutrition significantly affects the rate of conditioning and consequently carcass characteristics, chemical composition and fatty acid profile (Webb, 2006). Meat quality can be manipulated by a variety of nutritional interventions, many of which have been implemented successfully in feedlots world-wide (Webb, 2006). Although the positive effect of nutrition is usually greater in monogastric animals like chicken and swine, significant changes in meat quality have been reported in ruminants, particularly for fatty acid profile. Manipulation of meat quality through nutrition is more practical and cost effective compared to new breeding strategies, management techniques or post mortem technologies used to improve meat quality (Webb et al., 2006). Post mortem techniques (e.g. cold storage or retiring excess carcass fat) to improve meat quality are often time consuming and extremely expensive compared to appropriate nutritional strategies (Webb, 2006).

The effects of nutrition on meat quality are more significant in terms of carcass characteristics and chemical composition. On the other hand, the effects of nutrition on fatty acid profile are generally small but significant. The effects of nutrition on fatty acid profile, although small, are often very important in terms of nutritive value, colour of the product, quality or consistency of fat (Webb, 2006).

Addition of yeast (Saccharomyces cerevisae) to the diet of ruminants has been explored by researchers (Pereira et al., 2001). The use of cultures, such as Saccharomyces cerevisae or its extracts, can improve weight gain, as a result of the response to increased dry matter intake (Wallace, 1994). Yeasts, especially Saccharomyces cerevisae, have been used in animal diets for several decades and are considered sources of high-quality proteins and B-complex vitamins, selenium and zinc (Queiroz et al., 2004).

Carcass Characteristics

The animals used to compare nutrition had similar age (20 to 22 months) and they were finished in feedlots. The treatments used to compare nutrition were: HAB (Bermuda grass hay+corn concentrate), SOS (sorghum silage+corn concentrate), HAY (Bermuda grass hay+corn concentrate+ yeast), SIY (sorghum silage+corn concentrate+yeast), COR (corn silage+corn concentrate), LIN (corn silage+linseed concentrate) and SOY (corn silage+soybean concentrate).

Final weight (FEW) is high for all treatments. This elevated final live weight can be explained by the genotype used in these experiments (ZebuxEuropean) (Prado et al., 2008a). A high slaughter weight is necessary in order to achieve satisfactory (fat thickness) finishing levels.

Hot carcass weight (HCW), hot carcass dressing (HCD), conformation (CON), carcass length (CAL), leg length (LEL) and texture (TEX) present variation around 15% due to nutrition.

Carcass conformation represents muscle development in the anterior and especially the posterior carcass regions (Muller, 1980).

Carcass conformation is positively correlated with several characteristics that express muscle quality, such as CAL, LEL, Longissimus muscle area (LMA) and fat thickness (FAT).

FAT ranges between 3.01 mm to 5.70 mm. Animals that receive corn silage+corn concentrate achieve higher values than animals that receive Bermuda grass hay +corn concentrate+yeast. This is because corn concentrate is more energetic. In this way, the excess energy provided by corn concentrate is stored as subcutaneous fat (Prado et al., 2008a).

Ito et al. (2009) found high values for LMA. This parameter changed from 64.7 to 92.2 cm2. In general, authors find an average of 65.0 cm2 for this parameter.

Chemical Composition

Moisture percentage has 2% of variation among different nutritional treatments. It ranges between 73.2 to 75.0% and thus moisture percentage is little influenced by nutrition.

Ash percentage ranges between 0.90 to 1.10%. HAB and SIY have the highest value and the lowest value is for HAY. Prado et al. (2008a) suggest that ash percentage has little variation as a function of nutrition.

Crude protein percentage has around 21% of variation due to nutrition. LIN shows 19.9% and SOS presents 24.1%. Prado et al. (2008a) suggest that the lowest value for LIN is because it is a concentrate with high levels of oil and it does not have high levels of crude protein. On the other hand, SOS has high levels of crude protein. This may be the reason for the high percentage of crude protein in Longissimus muscle of animals on SOS treatment.

The variation in total lipids percentage in Longissimus muscle is 110% between the diets HAY and SOY. Animals fed with the HAY diet (1.30%) present the lowest value for total lipids. However, animals fed with the SOY diet (2.73%) present the highest percentage of total lipids in the Longissimus muscle.

Total cholesterol ranges between 21.5 to 55.4 mg/100 g of muscle. The treatments that receive the diets HAB, SOS, HAY and SIY produce lower values than the treatments that receive the diets COR, LIN and SOY. The values obtained for the diets HAB, SOS, HAY and SIY can be considered low in comparison to cholesterol levels cited in the literature. However, Silva et al. (2002) found levels of 18.4 mg/

100 g meat from crossbred heifers (Nellore vs. Simmental) fed with corn and yeast addition. Perhaps these low values are associated with a method that does not extract all the cholesterol in muscle.

These data show that there is an influence of nutrition on fatty acid profile. More than 60% of fatty acids percentage had variation due to nutrition. The principal class of fatty acids that suffers with nutrition is PUFA. The lower percentage is for COR treatment and the highest percentage is for HAB treatment. SOY is a concentrate with high levels of monounsaturated fatty acids (Prado et al., 2008d) and HAB is a roughage with high levels of polyunsaturated fatty acids (Maggioni et al., 2009). The fatty acid profile of the diet is important to know as the factor that influences fatty acid profile in Longissimus muscle.

The sum of the percentages of the fatty acids 14:0 and 16:0 found in Longissimus muscle of cattle is lower in diet SOS (26.4). Thus, if the objective is to obtain meat with low percentages of hypercholesterolemic fatty acids, the diet of sorghum silage with corn concentrate and yeast can be an alternative.

n-6 and n-3 percentages change due to nutrition. n-6 ranges between 4.90 to 10.7%. COR (4.90%), SOY (6.07%) and LIN (7.11%) show the lowest percentages. The diets HAB (10.7%), HAY (10.5%), SIL (9.41%) and SLL (8.87%) produce the highest percentages of n-6. Thus, nutrition influences n-6 percentage in Longissimus muscle of cattle. In the same way, n-3 has a variation due to nutrition and ranges between 0.83 to 2.12%.

Animals fed with HAB (0.28), HAY (0.25), SOS (0.22), SIY (0.22) and LIN (0.20) present higher values of PUFA/SFA in relation to animals fed with SOY (0.16) and COR (0.12). This characteristic changes around 110%. The English Department of Health (HSMO, 1994) recommends a ratio of about 0.45.

The best ratio of n-6/n-3 is found in animals that received the diets LIN (4.83), HAB (5.07) and SOY (5.36). However, no diet presents the ideal ratio recommended by the English Department of Health (HSMO, 1994) of 4.0 or less.

Finishing Systems

As the ethanol and renewable fuel industries grow, the availability of distiller's grains and other by-products continues to increase. These by-products are valuable feedstuffs for ruminants because of their high protein content and also their high fiber content. However, as more acres are dedicated to corn production, less pasture will be available for grazing livestock and less corn may be available for use as feed. As such, supplementing pasture-fed cattle with distillers grains may be an option for some producers to utilize.

Feedlot has been a cost-effective alternative for raising beef cattle in regions where either the price of grassland or dietary components are inflating operating costs. Consequently, both conditions require the use of intensive systems to produce high-quality meat. Presently, the use of cereal grains (such as corn) has been the main source of energy in finishing diets, but oils and fats can also be used as alternative components (Prado et al., 2008b).

Consumers demand beef with most of the fat removed but with a quality grade to ensure high palatability. Producers must continue to find new systems for producing cattle that reach a quality grade and still have minimal carcass fat (Camfield et al., 1999). Alternative production systems that include the use of roughages could range from pasturing cattle, pasturing cattle with a limited amount of concentrates, pasturing cattle and then feeding them concentrates for a short period of time in a feedlot, or feeding roughages to cattle while they are in the feedlot (Schaake et al., 1993).

Carcasses of roughage-fed beef are lighter and have less marbling and lower quality grades but have higher cutability than carcasses of grain-fed bulls (Prado et al., 2008b; c). Roughage-fed cattle are generally older than their grain-finished counterparts when they reach a choice quality grade. Carcass cutability varies (Schaake et al., 1993) and European breeds normally have higher yields of lean edible product than Bos indicus breeds. Although studies have shown compositional differences among breed types, it is important to understand how various cattle types can be optimally used to produce lean and high-quality beef (May et al., 1992). Cattle growth type and end point of production are critical in order for producers to successfully target carcass characteristics (Wheeler et al., 1996).

Carcass Characteristics

The animals used to compare finishing systems had a similar age (20 to 27 mo). The treatments used were a pasture system with supplementation and a feedlot. In the pasture system two genetic groups were compared: Puruna and NellorexAberdeen Angus. In the feedlot system the same genetic groups were used to compare the influence of finishing systems. In the same way, HCW has a variation due to the finishing systems. Bulls finished in a feedlot show the highest average for HCW. This is due to the high average that these animals show for FWE.

CON has a variation of 68.3%. Animals finished in pasture systems present a better average than animals finished in a feedlot.

The principal effect observed by different finishing systems is FAT and MAR. Animals finished in a feedlot show higher averages for FAT and MAR than animals finished on pasture (Padre et al., 2006). Prado et al. (2008c) suggest that the higher value found for FAT and MAR in animals finished in a feedlot is because these animals receive high amounts of concentrate with high energy, and the excess energy is synthesized into fat that is stored in subcutaneous muscle. MAR is synthesized after the FAT. When an animal has stored high quantities of FAT, synthesis of MAR starts (Prado et al., 2008b).

Chemical Composition

In this review, the variations for moisture, ash and crude protein percentages are lower than 6% due to the different finishing systems.

The different finishing systems seem to influence percentage of total lipids. Animals finished in a feedlot present higher percentages of total lipids than animals finished in a pasture system. Total lipids percentage ranges between 0.95 to 2.01%; these percentages are considered low due to the age of these animals (around 24 mo). The oldest animals present a high percentage of total lipids (Prado et al., 2009a).

On the other hand, the variation for total cholesterol is lower than for total lipids. It ranges between 41.1 to 48.4 mg/100 g of muscle. There is 15% of variation in total cholesterol due to finishing systems; animals finished in a pasture system show high values for total cholesterol.

Fatty Acid Profile

Finishing systems have an influence on variation of fatty acid profile. Variation can be observed for 18:3 n-6 (0.01 to 0.11%), 18:3 n-3 (0.25 to 1.14%), 18:2 c-9 t-11 (0.17 to 0.58%), 20:4 n-6 (1.49 to 2.54%), 20:5 n-3 (0.12 to 0.42%), 22:5 n-3 (0.15 to 0.98%) and 22:6 n-3 (0.06 to 0.36%). Almost 50% of fatty acids are highly influenced by the finishing system. This is because of the diet given to animals and the physiological alterations due to animal movement. Animals kept in a pasture system have to walk more in looking for feed than animals kept in a feedlot. This acts in the animal organism to change the metabolism so that more energy is expended in moving (Prado et al., 2008c).

The variation of the fatty acids 14:0 and 16:0 are 58 and 25%, respectively. Animals finished in a pasture system with supplementation present low percentages for 14:0 and 16:0 in comparison to animals finished in a feedlot. However, the variation in 18:0 is low (around 1%) between the finishing systems (Prado et al., 2008c).

The percentage of 18:3 n-3 is four times higher in Longissimus muscle of animals finished on pasture with supplementation in comparison with animals finished in a feedlot.

The higher percentage of this fatty acid found in Longissimus muscle of animals finished in a pasture system with supplementation also influences the percentage of 20:5 n-3. This occurs due to 18:3 n-3 being the precursor of 20:5 n-3 (Campbell, 1995). However, the higher percentage of 22:6 n-3 is observed in animals finished in a feedlot. This fatty acid also can be produced by the process that converts 18:3 n-3 into 22:6 n-3 (Campbell, 1995). However, the high percentage of this fatty acid can be due to its high percentage in the diet given to the animals finished in a feedlot.

The principal alteration can be observed in PUFA. Animals kept in a pasture system present with a higher percentage of polyunsaturated fatty acids than animals kept in a feedlot. Prado et al. (2008d) suggest that this difference is due to the fatty acid profile of pasture that is rich in polyunsaturated fatty acids. Silva et al. (2008), working with supplementation of Nellore steers, studied the effect of levels of supplementation in these animals and fatty acid profile was higher for PUFA in pasture (Brachiaria brizantha) than in the concentrate given to the animals.

In the same way, variation can be observed in n-6 and n3 percentages. These two classes of fatty acid are present in high percentages in Longissimus muscle of animals finished in a pasture system. This is because pasture has a greater percentage of these fatty acids than the concentrate given in the feedlot (Prado et al., 2008b).

The better ratios of PUFA/SFA and n-6/n-3 are found in Longissimus muscle of animals finished in pasture systems. Animals kept on pasture present a PUFA/SFA near that recommended by the

Department of Health (1994) of more than 0.45, and n-6/n-3 of less than 4.0 found in animals kept in pasture systems is ideal according to the recommendation of the Department of Health (1994),. This ratio appears to have positive effects on membrane fluidity, gene expression, cytokine formation, level and composition of serum lipids, and modulation of immune responses (Simopoulos, 1996). The two different genetic groups kept in a pasture system have a ratio of 2.35 (Puruna) and 2.38 (NellorexAberdeen Angus). The same breeds finished in a feedlot present high ratios (6.92 (Puruna) and 6.70 (NellorexAberdeen Angus)).

Gender

Nowadays, the food-industry prefers to buy steers because they have carcasses with higher fat deposits as already indicated by fat thickness and marbling. On the other hand, ranchers prefer to raise bulls because they grow faster.

Castration alters the growth rate and carcass characteristics due to modifications of hormonal status (Hunt et al., 1991). Otherwise, the higher growth rate of bulls may be caused by the gradual increase of hormonal secretion throughout their growth period. It seems that this higher growth rate is caused by anabolic hormones produced by the testicles (Lee et al., 1990). Additionally, some qualitative characteristics, such as protein and fat proportions are influenced by steroid hormones (Gariepy et al., 1990).

The management of bulls in an early-weaning system using high-energy diets enables early intramuscular deposits of fat, fast and efficient animal growth, and the production of lean carcasses of high quality (Schoonmaker et al., 1999). The ability of bulls finished in a feedlot to deposit intramuscular fat may be hindered because hormone secretion is high and deposition of muscle is more likely

Cull cows from dairy herds are of considerable economic importance to the producer and beef food chain (Jurie et al., 2007). However, producers want to maximize the value of cull cows to their business. Ensuring that animals are finished and the requirements of the processors are achieved is essential to optimize farm gate returns for these cattle (Lee et al., 2009).

The value of culls cow is highly sensitive to specific market conditions but maximizing returns from dairy cull cows is also sensitive to the costs associated with feeding to achieve slaughter weight (Lee et al., 2009).

Dairy farmers have three options when the fertility of dairy cows decline: i) remove cows from the milking herd and finishing for slaughter; ii) sell cull cows at salvage value for milking or to the slaughter house; iii) carry out extended lactations with rebreeding to facilitate calving at 18 or 24 month intervals (Auldist et al., 2007).

Carcass Characteristics

The age of the crossbred animals (1/2 Zebux1/2 European) used for this review on carcass characteristics were: bulls and steers (35 mo old), female cattle (102 mo old and 18 mo old). All animals were finished in feedlot.

Bulls show the highest FWE than other genders (steers and female cattle). This highest growth of bulls in comparison with steers or even female cattle seems to be due to the higher production of anabolic hormones by the testicles (Lee et al., 1990). There is a big variation among animals of different physiological condition. FWE ranges between 317 to 578 kg. Heifers have the lowest value. Marques et al. (2006) suggest that the low FWE is due to this gender having less muscle deposition.

HCW is influenced by FWE (Prado et al., 2009a). Thus, heifers show low values for HCW (171 kg) and bulls present high values (307 kg). Steers (269 kg) and cull cows (245 kg) present intermediate values.

CAL has more variation for heifers. This gender presents the lowest value for carcass length due its size; heifers are smaller than bulls, steers and cull cows (Prado et al., 2009a). In the same way, LEL and CUT have low values for heifers (68.2 cm; 21.0 cm). Bulls (78.9 cm; 25.3 cm) and steers (78.0 cm; 24.0 cm) do not show variation.

FAT is one of the principal parameters that is affected by different physiological conditions. Cull cows (4.72 mm) and heifers (4.00 mm) present higher values than bulls (1.75 mm) and steers (2.81 mm). This is because the female deposits more fat than the male (Prado et al., 2009a). Steers have higher FAT than bulls because of the lower production of testosterone by steers (Prado et al., 2009a).

LMA has variation among animals of different physiological condition. It ranges between 50.3 to 65.1 cm2. Bulls have a high value due to their capacity to deposit muscle. Cull cows present a low value because of their deficient deposition of muscle; they deposit more fat than muscle (Lee et al., 2009).

COL and TEX present little variation. However, MAR shows variation among bulls, steers, cull cows and heifers. Cull cows (5.40 points) and heifers (3.70 points) present more marbling due to their capacity to deposit fat. Marbling is the last fat to be deposited and females have genes that control the deposition efficiently (Lee et al., 2009). Bulls have a low value for MAR of 1.50 points. For steers, this value is 2.88 points.

Chemical Composition

The ages of the crossbred animals (1/2 Zebux1/2 European) used for this review on chemical composition were: bulls and steers (35 mo old), female cattle (60 mo old and 20 mo old). All animals were finished in a feedlot.

Moisture percentage has variation among bulls (75.0%), steers (72.8%), cull cows (71.4%) and heifers (74.4%). This variation between bulls and cull cows is due to the high percentage of total lipids present in their Longissimus muscle. Prado et al. (2008b) suggest that the variation on moisture percentage is due to the variation in total lipids.

Usually, ash percentage does not change due to genetic groups, nutrition and finishing systems (Prado et al., 2008a; b; c; d). However, gender is a factor that has influence on ash percentage. Cull cows (1.40%) present a higher ash percentage than other categories (bulls (1.08%), steers (1.06%) and heifers (1.13%)). The high value found for cull cows is because these animals were slaughtered at 60 mo old and the other categories were slaughtered earlier (30 mo old) (Minchin et al., 2009).

Crude protein percentage presents variation between male and female. Bulls (24.3%) and steers (23.3%) do not have variation. In the same way, cull cows (21.1%) and heifers (21.2%) do not have variation, but males have higher crude protein percentage than females.

The parameter of chemical composition that is more influenced by gender is percentage of total lipids. Cull cows (7.10%) have more total lipids percentage than bulls (0.91%), steers (1.46%) and heifers (2.42%). Minchin et al. (2009) suggest that the higher percentage for total lipids in cows is due to their high deposition of fat. The low percentage of total lipids in bulls is explained by testosterone; this hormone is related to the higher capacity for muscle growth in bulls and the lower capacity for fat deposition (Rule et al., 1997).

Total cholesterol has variation among bulls (62.3 mg/100 g of muscle), steers (52.0 mg/100 g of muscle) and heifers (49.7 mg/100 g of muscle). The difference between bulls and steers is explained by changes in the cellular structure of the muscle (Rule et al., 1997). Thus, bulls present with more total cholesterol content than other animal categories.

The percentages of saturated fatty acids C14:0 and C16:0 are higher in females than males, but there is no variation in percentage of these fatty acids between bulls and steers or between cull cows and heifers. However, C18:0, an abundant saturated fatty acid, is higher in males than females. Cull cows present the lowest percentage of this fatty acid. The low percentage present in Longissimus muscle of this animal category is partly explained by biohydrogenation reactions in the rumen (Tamminga and Doreau, 1991). The majority of adipose deposits in animal tissues are synthesized by lipogenesis because ruminant diets are poor in fat components; fatty acids are elongated up to 18:0 and converted into 18:1 by desaturation (Rule et al., 1997). The deposition of 18:1 is also increased while 18:0 is reduced during the deposition period; this could explain the higher levels of 18:1 in steer muscles.

C18:2 n-6 presents variation among gender. Cull cows (1.87%) have the lowest percentage of this fatty acid. Bulls (12.6%) have the highest percentage of C18:2 n-6. There is a high variation (674%) between these classes. In the same way, C20:4 n-6 has the highest percentage for bulls (3.14%) and the lowest percentage for heifers (0.93%). This demonstrates that males have a greater percentage of essential fatty acids than females.

SFA does not present variation among gender. MUFA is influenced by this characteristic and bulls present the lower value. On the other hand, cull cows (47.0%) have a high percentage of MUFA. SFA have been implicated in diseases associated with modern life, especially in developed countries; these include various cancers and especially coronary heart disease (Wood et al., 2003). MUFA is a class of fatty acids that have a positive influence on health. C18:1 n-9 (oleic acid) increases the level of HDL-cholesterol (High Density Lipoprotein) and reduces the level of LDL-cholesterol (Low Density Lipoprotein) in human blood (Katan et al., 1994) and studies in humans have demonstrated a strong relationship of LDL-cholesterol to HDL-cholesterol and higher risk of cardiovascular disease (Kwiterovich, 1997).

Bulls present a high value for n-6 of 16.7%; steers present 10.1%, cull cows present 2.97% and heifers present 4.37%. There is variation for this fatty acid due to gender. In the same way, bulls have a high n-3 percentage of 4.03%. Steers present 2.70%, cull cows present 1.77% and heifers 0.79%.

The Department of Health (1994) recommends that PUFA/SFA should be higher than 0.45. In this case, only bulls present the ideal ratio, namely 0.46. The other categories present low values for this ratio. Meat cattle usually have PUFA/SFA lower than 0.45 so, meat has been implicated as a cause of the imbalanced fatty acid intake of today's consumers (Wood et al., 2003). For this reason, ways to improve PUFA/SFA during meat production are required. Researchers have focused on the type of PUFA and the balance in the diet between n-3 formed from linolenic acid (C18:3) and n-6 formed from linoleic acid (C18:2) (Williams, 2000).

n-6/n-3 presents better values for steers (3.71) and cull cows (1.70). The other genders obtained high values. The recommendation by the Department of Health (1994) is less than 4.00. The n-6/n-3 value is also a risk factor in cancers and coronary heart disease, especially in the formation of blood clots leading to a heart attack (Enser, 2001).

Implications

The carcass characteristics, chemical composition and fatty acid profile are influenced by genetic groups, nutrition, finishing systems and gender. Bulls reach final weight faster than steers and females. Genetic groups that contain zebu genes in their composition present high hot carcass dressing. However, the Longissimus muscle of genetic groups with zebu genes present low marbling. The crude protein content is altered by nutrition. High crude protein percentage is observed in animals fed with sorghum silage. The Longissimus muscle of cull cows presents a high percentage of total lipids in relation to other genders. Nutrition also influences the fatty acid profile. Animals fed with Bermuda grass hay+corn concentrate present high values for polyunsaturated fatty acids, n-6, and n-3 fatty acids, and better ratios of PUFA/SFA and n-6/n-3. Better ratios are also found in the Longissimus muscle of animals finished on pasture with supplementation compared with animals finished in feedlot. Females present inferior meat quality in relation to the percentage of hypercholesterolemic fatty acids (14:0 and 16:0), lower percentage of essential fatty acids and less PUFA.

5

Comparison of Genetic and Multiple Trait Animal Models

Introduction

Artificial insemination plays an important role in animal breeding by allowing greater utilization of genetically superior sires. The opportunity for genetic improvement of male fertility traits has been previously. However, appropriate statistical methods for analysis of semen traits in pigs has not been extensively studied.

In previous studies, this type of data was analyzed by multiple trait methods choosing the most important time points as separate traits. Because of the number of potential observations over the lifetime of a boar, it would be difficult to thoroughly analyse this type of data due to computational limits. Semen data have also been analysed similarly to growth curves ignoring genetic effects, or were considered simple repeated measurements ignoring time dependency.

In many cases, the assumption of a univariate repeated model is not appropriate while a full multivariate model with the number of traits equal to the number of ages would result in a highly over-parameterized analysis (Meyer and Hill, 1997). Random regression models (RRM RRM Radio Resource Management (GSM/UMTS)

RRM Resilient Risk Management) developed by Meyer (1998) have been extensively applied to the test-day model analysis of milk yield of dairy cattle and have also been fitted to weight data of pigs. RRM provide a method for analysing independent components of variation that reveal specific patterns of change over time.

Evidence has been reported that changes in animal performance with increasing age are influenced by genetic factors. Animal breeders are interested in genetic parameters that describe the change of traits over time. Analysis of these changes can be undertaken using repeatability (Henderson, 1984), multiple trait (Reents et al., 1995) or random regression models. Random regression allows for the calculation of (co) variances at every age (Meyer, 1998). Multiple trait animal models have traditionally been used for traits measured over time by defining observations at distinct ages as different traits. However, computational requirements need to explain the number of traits equal to the number of ages.

Accordingly, records collected over ages are often analysed as repeated measurements or as different traits that are separated by specific intervals. However, it is the object of interest how much RRM is better than multi-trait analysis. The objective was to compare the (co) variance of total sperm cells (TSC TSC Thestreet.com (stock symbol)

Materials and Methods

Data Source

Records of total sperm cells per ejaculate (n = 19,629) from 834 boars were provided by Smithfield Premium Genetics (Rose Hill, NC). One thousand seven hundred and thirty six individuals were included in the pedigree pedigree

Record of ancestry or purity of breed. Pedigrees of domesticated animals are maintained by governmental or private record associations or breed organizations in many countries. file. Boars represented three breeds and were housed in AI stations. Each AI station was similar in number of boars of each breed and management. Thirty-four collectors recorded these data over 5 years (1998 to 2002) with approximately one-half of all records in 2000. Data were distributed evenly across seasons.

An instrument used for making measurements of light, or electromagnetic radiation, in the visible range. In general, photometers may be divided into two classifications: laboratory photometers, which are usually fixed in position and yield results. Observations were removed when the number of data at a given age of boar classification time point was less than 10, or total sperm cells were missing, zero or less than zero. Differences between boar collection date and birth date were used to provide each record with a fixed age of boar classification in weeks for RRM. When a boar had two observations during one week of age the record closest to the whole week was utilized. For the multiple trait analyses, records were edited to include only records produced at 9, 12, 15, 18, 21, 24, and 27 months of age and were used as separate traits. Number of observations at 9, 12, 15, 18, 21, 24, and 27 months of age were 305, 413, 370, 306, 248, 200 and 109, respectively. Number of animals with valid records was 750. Frequency of records was highest at 12 months of age, decreasing gradually over time.

Multiple Trait Animal Model

Least square means were estimated for the fixed effects of breed and AI station, and differences within fixed effects were compared by least significant differences using the PDIFF option in SAS (1). A software company that specializes in data warehousing and decision support software based on the SAS System. Founded in 1976, SAS is one of the world's largest privately held software companies. The statistical model included fixed effects of year-season, breed, collector, and AI station.

Variance components for the multiple trait analyses were estimated by derivative free REML REML Restricted Maximum Likelihood (statistical) using MTDFREML (Boldman et al., 1995). Fixed effects for the model were year-season, breed, collector and AI station. Convergence was considered to have been reached when the variance of the -2 log likelihood in the simplex was less than 1x[10.sup.-9]. To obtain convergence, as well as standard errors for parameter estimates, the seven ages classifications of total sperm cells were evaluated in five trait analyses. Therefore, twenty-one combinations of five trait analyses were conducted. The results presented are the means for each parameter estimate, and standard deviations were considered as standard errors. A measure of the degree to which returns on two risky assets move in tandem. A positive covariance means that asset returns move together. A negative covariance means returns vary inversely. matrices, with order equal to the number of traits in the analysis, and A is Random regression model.

Random regression procedures are fully described in Oh et al. (2006). Parameters were estimated for total sperm cells by age of boar classification under a random regression model using DxMRR (Meyer, 1998). The analysis model included breed, collector and year-season as fixed effects; additive genetic effects, permanent environmental effect of boar and measurement error as random effects Random effects can refer to:

- Random effects estimator
- Random effect model.

Random regression models were fitted to evaluate all combinations of first-through seventh-order polynomial polynomial, mathematical expression which is a finite sum, each term being a constant times a product of one or more variables raised to powers. With only one variable the general form of a polynomial is $a_0x^n + a$ covariance functions for fixed age of boar classification, additive genetic and permanent environmental effects. This resulted in the evaluation of 343 models. Goodness of fit Goodness of fit means how well a statistical model fits a set of observations. Measures of goodness of fit typically summarize the discrepancy between observed values and the values expected under the model in question. Such measures can be used in statistical hypothesis testing, e. for models was tested using log likelihood value, Akaike's Information Criterion There are a number of statistics that can act as an information criterion. They include:

- Akaike's information criterion
- the Bayesian information criterion, also known as the Schwarz information criterion
- Hannan-Quinn information criterion.

AIC $= -2\log L + 2 \times p$

SC $= -2\log L + p \times \log(N-r(X))$

Where, p is the number of parameters estimated and r(X) is the rank of the coefficient matrix In linear algebra, the coefficient matrix refers to a matrix consisting of the coefficients of the variables in a set of linear equations. Example

In general, a system with m linear equations and n unknowns can be written as of fixed effects.

Multiple Trait Animal Model Analyses

Mean and standard deviation of total sperm cells increased with age. Coefficient of variation Coefficient of Variation. A measure of investment risk that defines risk as the standard deviation per unit of expected return. was the highest at 9 months of age and lowest at 27 months of age, however, there was not much difference among ages. Breed 3 showed significantly more total sperm cells than both breeds 1 and 2 only at 15, 21 and 24 months of age (p<0.05). At 9, 12 and 18 months of age, breed 3 had more total sperm cells than breed 1 but not breed 2 (p>0.05). Also, there was no breed effect observed at 27 months of age in the comparison of total sperm cells produced. Artificial insemination stations did not differ (p>0.05) except at 15 months of age. There was no AI station by age interaction (p>0.05).

Estimates of genetic variance for total sperm cells were lower for boars 9, 12 and 15 months of age. The genetic variance for total sperm cells was also much higher at 24 months as compared to all other ages, which may be due to sampling of records. Estimates of permanent environmental variance for total sperm cells were close to zero and did not differ across age classifications. Estimates of residual variance for total sperm cells were lower at 9, 12 and 27 months and higher at 15, 18 and 21 months. A significantly lower residual variance was observed for total sperm cells at 24 months. Again, this may be due to sampling of records. These results suggest that both genetic and residual variance for total sperm cells over age classifications are heterogeneous and possibly different traits.

Heritability estimates for total sperm cells were similar across age classifications with the exception of 24 months of age. Heritability of total sperm cells at 24 months of age was high because of high genetic variance and the low residual variance at that age. This may be due in part to selection of records for specific age points. Heritability estimates in this study were similar to those reported in the literature. Masek et al. (1977) estimated 0.24 as repeatability using two-factorial hierarchical analysis of variance.

Genetic and residual correlations between measures of total sperm cells at different ages averaged 0.64 and 0.30, respectively. Genetic correlations between adjacent ages were higher than those between more distant ages. Decreasing genetic correlations with increasing age may also be due to the limited amount of data and the selection of records to defined age ranges. Huisman et al. (2002) reported a similar observation in an evaluation of pig body weights.

Random Regression Model Analysis

Results from the random regression analysis are presented in Oh et al. (2006). In brief the random regression model, fitting 6th, 5th, and 7th order for fixed, additive genetic and permanent environmental effects showed the largest log likelihood value. This model was the 4th best fitting model based on AIC and the 52nd best fitting model based on SC. AIC showed best fit when, respectively, 6th, 4th, and 7th order fixed, additive genetic and permanent environmental effects were fitted. This was the 3rd best-fitting model based on log likelihood and 20th best-fitting model based on SC. Schwarz Criterion showed the best fit when 4th, 2nd, and 7th order polynomials were fitted for fixed, additive genetic and permanent environmental effects, respectively. This model was ranked with the 10th best fit by log likelihood and 2nd best-fitting model by AIC. Based on the conservative nature of SC and the relative ranking by the other criterion, this model may be the best overall fit. Heritability estimates for total sperm cells over weeks of age ranged from 0.27 to 0.48. Standard deviations tended to decrease from 33 weeks of age to about 45 weeks, maintained consistent intervals by 100 weeks of age and then increased rapidly. This increase in variance closely follows the numbers of total sperm cells records over age.

Comparison Between Models

Estimates of genetic parameters from both multiple trait and random regression methods would indicate that measures of total sperm cells at different ages are genetically different traits. The comparison of heritability estimates between the three best-fit models determined from random regression model analysis and the evaluation of seven ages by multiple trait animal model analyses. The results are similar except for the heritability at 24 months of age from the multiple trait method that had very high genetic variance. Other than 24 months, the results are consistent but it appears that the multiple trait method resulted in a higher estimate of heritability of total sperm cells at each age. These higher estimates may be due to the reduced amount of data used in the multiple trait method or possibly the age classifications selected. The ability to accurately estimate genetic correlations between different ages is reduced by limiting records to specific ages. Therefore, multiple trait methods that do not include all available data may not be the most appropriate method for analysing longitudinal data. However, this method may be improved with sufficient numbers of records at each age and if availability of computer resources allowed for more age classifications. Random regression analysis provides much more detail with regard to the changes of the variance components with age. Genetic correlations between total sperm cells at different ages were larger for adjacent ages. RRM with comparatively high order polynomials for fixed, additive genetic and permanent environmental effects provided the best fit.

These studies show there is an opportunity for genetic selection on semen traits. Estimates of genetic parameters would indicate that measures of total sperm cells at different ages are genetically different traits. However, the ability to accurately estimate genetic correlations between different ages is reduced if records are limited to specific ages. Therefore, multiple trait methods that do not allow for the inclusion of all available data may not be appropriate for analysing longitudinal data. This method may be appropriate with sufficient numbers of records at each age and availability of computer resources. Random regression methods are the most appropriate to analyse semen traits as they are longitudinal data measured over the lifetime of boars. Additional work is needed to understand the relative economic importance of semen traits in the development of breeding objectives.

Genetic Parameters and Annual Trends for Birth and Weaning Weights of a Northeastern Thai Indigenous Cattle Line

The Thai (or Tai) are the main ethnic group of Thailand and are part of the larger Tai ethnolinguistic peoples found in Thailand and adjacent countries in from within the country; utilize the adapted genotype genotype. Genetic makeup of an organism. The genotype determines the hereditary potentials and limitations of an individual.

Growth traits such as birth weight and weaning weight are of primary economical importance in cow-calf production system. They are known to be influenced by the direct genetic effect of the calf and the maternal genetic effect. The primary goal of animal breeders is to maximize the rate of genetic improvement. Knowledge of the nature and magnitude of population parameters (i.e. variance components and heritabilities) for a Northeastern Thai indigenous cattle line is needed to effectively design breeding programs and to estimate breeding values for traits of relevance to beef producers.

In order to achieve optimum progress in the selection program for cow-calf production system both direct and maternal genetic effect should be taken into account, especially if an antagonistic relationship between them exists. Apart from this, annual trends for calf weights should be monitored over time to check the validity of the predictions made and to investigate direction of genetic change and whether the selection strategies implemented could reach a selection limit or have unexpected other effects. Therefore the present study was conducted with the following objectives: 1) To estimate (co)variance components and genetic parameters for birth and weaning weight of a Northeastern Thai indigenous cattle line 2) To estimate direct and maternal genetic effect for birth and weaning weight of a Northeastern Thai indigenous cattle line and 3) To measure phenotypic and genetic trends for birth and weaning weight of a Northeastern Thai indigenous cattle line.

Materials and Methods

Animal Management

Records of birth and weaning weight of calves of a Northeastern Thai indigenous cattle line born between 1993 and 2004 were obtained from the Department of Livestock Development. The animals were raised in three stations (Chaiyapoom, Ubonrachatani and Udonthani Livestock Research and Breeding stations) which are located in Northeastern part of Thailand. Natural mating was used at a ratio of 1 bull per 25-30 cows and calving calving act of parturition in a bovine female, and presumably in any animal that bears a calf as its newborn. Calves were weighed within 24 hours after birth and stayed with their dams until weaning (approximate 7 months of age). All calves were identified at birth with tattoo, dehorned and treated against internal parasites and vaccinated for food and mouth disease at 3-4 month of age. The cattle in each station were allowed to graze on pasture during the day and were confined in large pens at night. Hay or silage silage or ensilage, succulent, moist feed made by storing a green crop in a silo. The crop most used for silage is corn; others are sorghum, sunflowers, legumes, and grass was fed ad lib An earlier sound card from Ad Lib, Inc., Quebec City, that, for a while, was the de facto standard for synthesized background music for computer games.

Dry lick mineral blocks and drinking water drinking water supply of water available to animals for drinking supplied via nipples, in troughs, dams, ponds and larger natural water sources; an insufficient supply leads to dehydration; it can be the source of infection, e.g. leptospirosis, salmonellosis, or of poisoning, e.g. were available ad lib. for each herd in the large pens. During the rainy season (June to December), the animals were kept on improved pasture (6 animals/ acre) without supplementation. The pasture were reestablished every three years by using ruzi grass (Brachilia ruziziensis) and verano stylo styloSa perennial legume adapted to grow in many environments. An alternative to alfalfa in tough environments. Called also *Stylosanthis gracilis*, Brazilian lucerne. (Stylosanthes spp.).

The calves were weaned four times a year (24 March, 5 June, 23 September and 5 December). Bulls and heifers were selected for replacement based on structural soundness and weight ratio using the expression of animal's performance for adjusted-weaning weight relative to contemporary group average. Bulls and heifers were selected if they had adjusted weaning weight ratio above the contemporary group average. The selected bulls from each station were assigned to mate with dams in all stations in order to establish genetic linkage Genetic linkage occurs when particular genetic loci or alleles for genes are inherited jointly. Genetic loci on the same chromosome are physically connected and tend to segregate together during meiosis, and are thus genetically *linked*.

Data Management

The pedigree and data file were edited to the same standard protocol. Only weaning age between 120 to 300 days were included. Records outside three standard deviations from the overall mean would be checked and eliminated using SAS (1). A software company that specializes in data warehousing and decision support software based on the SAS System. Founded in 1976, SAS is one of the world's largest privately held software companies. Other basic editing involved consistency checks for sex, age at weighing and age of dam including pedigree validation. For a small proportion of animals (584 out of 1,922 records), age of dam could not be obtained due to missing birth date of the dam and was replaced by the mean age of dam in the data set (4.8 years). This occurred mainly in the early years of the project for foundation cows with purchasing parent stock from the farmer. In order to determine the fixed effects to be included in the model, preliminary analyses were performed using the PROC (language) PROC-The job control language used in the Pick operating system.

MIXED in SAS (1996). If effects were notsignificant, they were eliminated from the model. Finally, the fixed effects taken into account included contemporary groups, defined as herd (station)-year-seasons of calving (3 seasons, i.e. Summer, March-June; Rainy, July-October and Winter, November-February) and sex of calves. Age of dam and age of animal were fitted as covariate.

The final data set comprised 1,922 and 1,489 records for birth and weaning weight in three stations, calves were by 48 different sires and 584 different dams. Approximately 80% of dams had more than 2 calves and 40% had their own birth and weaning weight recorded.

Estimate (co) Variance Components, Genetic Parameters and Breeding Value

A bivariate analysis was used for estimating variance and

covariance components by the average information restricted maximum likelihood (AI-REML) and fitting an animal model using ASREML (Gilmour et al., 2002). The preliminary analyses were conducted for direct-maternal genetic correlation for both traits however the results were close to zero. Therefore these effects were ignored in this study. The models were as follows:

Phenotypic and genotypic genotypicselection of breeding stock on the basis of known inherited characteristics.

Breeding values for birth and weaning weight for individual animals were estimated utilizing all available pedigree and performance information. Average estimated breeding value and maternal breeding value of the year 1993 were set to be zero as a base group. The genetic trend was calculated for each trait by regressing the average estimated breeding values and maternal breeding values that deviated from the base group on birth year of calves. The phenotypic trend was calculated for each trait by regressing the adjusted weight on birth year of calves. Adjusted weights were obtained from a least square analysis.

Results and Discussion

Genetic Parameters

The estimate of variance component and genetic parameters with their standard error are present. For birth weight, both direct and maternal genetic effects played more an important role on birth weight than that of non-genetic, i.e. maternal permanent environmental effect. The estimates of direct and maternal heritabilities were moderate to low, explaining 40% and 14% of the phenotypic variance respectively. The maternal heritability amounted to 35% of that due to direct heritability. Birth weight was affected by maternal permanent environmental effect to a less extent, explaining 4% of the phenotypic variance. As found in other breeds, Meyer (1993) and Robinson (1996) reported that direct heritability, maternal heritability and maternal permanent environment effect were 0.43, 0.10 and 0.09 for Hereford and 0.35, 0.08 and 0.05 for Angus, respectively. The results were agreement with this analysis, identifying birth weight as highly to moderately heritable trait with the maternal heritability amounting to about 23% of that due to direct heritability.

For weaning weight, the estimates of direct heritability and maternal permanent environmental effect played an equally important role, explaining 27% and 23% of the phenotypic variance. The direct heritability was found to be moderate. The maternal permanent environmental effect was the main factor determining weaning weight when compared to that of birth weight. In contrast, the maternal heritability which generally assumed to express the variation in milk production potential indicated a low variation for mothering ability, explaining 5% of the phenotypic variance. The maternal heritability amounted to 18% of that due to direct heritability. Similar results were observed by Meyer (1992), Meyer (1993) and Meyer et al. (1993) who reported that maternal permanent environmental effects in Hereford, Charolais were consistently more important than maternal genetic effect. The estimate of direct heritability in this study was well in agreement with the range of 0.21 to 0.30. Weaning weight was found to be moderately heritable trait and maternal heritability amounted to be average 14% of that due to direct heritability. However, the estimates were much higher than the estimate of direct heritability reported for Korean cattle.

In a Northeastern Thai indigenous cattle line, the direct heritability for weaning weight was considered moderately heritable trait. It was therefore possible for genetic improvement through selection in this population. The maternal heritability was low and played less an important role than direct heritability which indicated that the improvement of weaning weight should be based on direct genetic effect.

Phenotypic and Genetic Correlations

The estimate of phenotypic, direct genetic, maternal genetic, maternal permanent environmental and residual correlations between birth and weaning weight are present. The phenotypic and direct genetic correlations were moderately to highly positive, 0.48 and 0.65 respectively. The result indicated that calves with high performance at weaning tended to be heavier at birth. For maternal effects, i.e. the maternal genetic and maternal permanent environmental correlations were found to be unity to highly positive. The residual correlation was lowly positive (0.25), which implied that the relationships between prenatal and postnatal postnatal /post ·na ·tal/ (-na´t'l) occurring after birth, with reference to the newborn or occurring after birth, especially in the period immediately after birth environment were only slightly positive. The surrounding environment experienced by a calf in uterus had little in common with the environment experienced by a calf after birth. On the whole, except for maternal effects, levels of these correlations were comparable to other breeds. As observed by many authors, allowing for a model with the direct-maternal genetic correlation yielded a large negative estimate for this parameter. The increasing estimate of both direct and maternal heritabilities dramatically augmented the likelihood significantly. The large negative estimate of direct-maternal genetic covariances did not reflect a marked adverse genetic correlation between growth and maternal performance. The negative values were associated with the management practice or environmentally induced negative dam-offspring covariances. For this study, allowing for a model with the direct-maternal covariances included resulted in slightly higher heritabilities for the direct as well as maternal heritability, 0.28 and 0.06 respectively. The direct-maternal genetic correlation was-0.36. However, the model did not augment the likelihood significantly when compared to that of excluded the two covariances. Therefore, the model with the direct-maternal covariance included was ignored from this study. Considerably more data across generations is required to investigate the negative covariance further.

Phenotypic and Genetic Trends

The phenotypic and genetic trends for estimated direct breeding values and maternal breeding values for birth and weaning weight were present.

The phenotypic trend of birth weight was positive, 0.18 kg/year (p<0.05). Ranges of least square means for birth weight were ranged from 14.75 to 17.68 kg. In 1999, birth weight were dramatically increased and gradually decreased thereafter.

In contrast to birth weight, the phenotypic trend for weaning weight was found to be-1.36 kg/year (p>0.05). Ranges of least square means for weaning weight (adjusted to 200 day weight) were from 77.40 to 105.86 kg, a reflection of variation in the environment and management over the years. The adjusted weaning weight was gradually decreased from 1994 to 1998. In 1999, weaning weight was dramatically increased and thereafter decreased to 79.06 kg in the last year. The reduction in mean weaning weight for calves born in the last five years (2000 to 2004) were mainly due to a larger proportion of calves selected in previous years in order to increase the population. Some bulls and heifers were selected when they had adjusted weaning weight ratio below the contemporary group average. Therefore, to improve phenotypic performance for weaning weight of the herd, improved management strategies are also needed. Changes in management such as pasture improvement, grazing strategies and culling procedures needed to be measured and monitored in order to evaluate the benefit of the change (Allen, 2002). Estimated breeding values for birth weight have increased since 1994. The averages of estimated breeding values for birth weight ranged from-0.60 to 0.14 kg and the genetic trend averaged 0.04 kg/year (p<0.05). The averages of estimated breeding values for weaning weight ranged from-2.96 to 1.21 kg and the genetic trend was positive, 0.32 kg/year (p<0.05).

Maternal breeding values for birth weight were also increased from the year 1994. The averages of maternal breeding values for birth weight ranged from-0.36 to 0.14 kg. The genetic trend was positive, 0.01 kg/year (p>0.05). The averages of maternal breeding value for weaning weight were ranged-1.49 to 0.55 kg. The genetic trend was positive, 0.03 kg/year (p>0.05).

The estimates of genetic parameters obtained in this study indicated that if genetic improvement through selection for weaning weight of a Northeastern Thai indigenous cattle line is desired, substantial genetic progress can be achieved. Estimated breeding values and maternal breeding values is another tool for cattle breeders to generate animals that are suitable for their cow-calf production system. Similarly, breeders should use various management practices to help achieve this aim as well. Genetic and management are complimentary and both require measurements to achieve results. Estimated breeding values could be used not only to select superior animals genetically for the primary traits of interest, but also to monitor secondary traits that are of little current interest but might become more important later on. In this study, selection for higher weaning weight has caused an increase in birth weight. Therefore, while herds are not experiencing calving difficulty at present, monitoring birth weight while selecting for higher weaning weight is a good insurance policy..

Maintaining Crossbred Populations of Dairy Cattle in the Tropics

Under nearly all conditions of climate and management throughout the world, except the very worst, the milk yield of a local cattle breed can be increased by crossing it with an improved dairy or dual-purpose (milkmeat) breed. Thus, as countries with a temperate climate develop their dairy industries they inevitably replace their local breeds by such improved breeds, either by grading up (displacement crossing) or by importing both cows and bulls. The same applies to developed countries in subtropical regions, such as Israel, South Africa, southern Australia and the southern United States.

As long as the European dairy cattle are well fed, protected from disease, shielded from direct sun and given a chance to cool down at night and in the winter, they do not appear to be seriously inconvenienced by high environmental temperatures. It may be difficult to get cows in calf during the hot wet season and there may be a seasonal decline in yield but in general a European dairy breed is the most economic type for these regions. The same applies to those tropical regions in which the climate is moderated by altitude, as in many parts of east and south-central Africa, or by proximity to the sea, as in the islands of the Caribbean and the South Pacific.

However, there comes a point as we approach the deep tropics where purebred European cattle require more elaborate management than is possible under the existing economic and technological conditions. Many trials in east Africa, India, southeast Asia and Central America have shown that crossbreds between local breeds and European dairy breeds not only produce more milk than the local breed but also more than the pure European. Naturally, if the first cross is half way between the two breeds and the backcross to the European breed is half way between halfbred and purebred then the genetic situation is purely additive and grading up to the European breed is indicated. This article is concerned with the case where optimum performance is given by some intermediate between the two breeds, and will discuss breeding schemes which can be used to maintain this hybrid vigour.

Maintaining a Crossbred Population

There are three ways of mating the F_1 females after the initial cross. They can be mated to sires of the local breed or of the improved breed or they can be mated to F_1 sires. On an experimental basis, all three matings should be made. This will make possible a comparison between the F_1 and F_2 and between halfbreds and three-quarterbreds. Hopefully, purebreds will also be present for comparison.

If the halfbred or threequarter is clearly superior then *inter se* breeding within this level of exotic blood should begin at once. There is no point in trying further to refine the exact proportions which may be optimal. There will be so much variation within grades that only an approximation is necessary and selection can at once be based on performance rather than on proportion of blood. Because the well-known Santa Gertrudis beef breed was based on 3/8 zebu and 5/8 British, there has been a tendency to extol this as the ideal proportion and it has been followed in several other new breeds; e.g., the Bonsmara of South Africa and the Jersind and Brown Sind of Allahabad. This is pure formalism. In practice there are three principal ways of maintaining a crossbred population, whether it starts from an F_1 or a backcross:

1. The crosses may be bred *inter se* to produce a new breed which will have, at least initially, either 50 percent or 75 percent of the blood of the exotic breed.

2. The crossbred bulls may be used to mate with, and eventually to grade up, the local cows. The proportion of exotic blood will thus rise gradually to reach the limiting values of 50 percent, or 75 percent when the displacement is complete and a new breed is formed.

3. The F_1 cows can be bred to bulls of the two pure breeds in alternate generations in a crisscrossing system. When equilibrium is reached the cows will alternate in successive generations between 66.6 and 33.3 percent exotic blood.

I.L. Mason is Animal Breeding Officer, Animal Production and Health Division, FAO, Rome.

In systems (1) and (2) 50 percent of the hybrid vigour of the F_1 is maintained in the new breed. System (3) retains two thirds of the hybrid vigour. On theoretical grounds this would be the system to use if a large decline in performance from F_1 to F_2 indicated the presence of hybrid vigour. However, important criteria are: which system can be operated in practice and which offers the best scope for a selection programme to follow the initial crosses? Effective genetic improvement in milk yield depends firstly on an efficient milk-recording system, and secondly on a bull progeny testing scheme based on the results of this recording.

If such a system is available it can be applied at once to the new breed. The breed formation will have to be made on a scale sufficient to provide opportunity for progency testing several bulls each year. This cannot be restricted to a single small government herd. Within a single experimental herd milk recording is easy enough but cows may be insufficient in number to allow an adequate progeny test of several bulls and a choice between them. If no milk-recording system is available then it may be better to exploit the improvement programme in another country (or possibly in a more developed part of the same country) by importing improved bulls or their semen and using a crisscrossing or rotational breeding system.

Formation of New Breeds

The activities of breeders and breed societies in western Europe and North America during the last century have taught us to think of breeds as closed population of nearly identical animals. From the present point of view the important characteristic of a breed is that it is self-contained. It generates its own supply of bulls and does not have to rely on an outside source of bulls or semen.

Because of the aura from the past and the emphasis on uniformity there has been a reluctance to take the step of attempting to form a new breed. People are afraid that breeding from crossbreds will lead to excessive variation, and therefore that formation of a new breed is only possible if immense resources of animals are available. In fact, the increase in variation on breeding an F_2 generation has been much exaggerated. To begin with, it is only the genetic variation which is increased, and with characters of low heritability this is only a small part of the total variation. Secondly, the obvious effects are due to segregation of colour, pattern and morphological characters (e.g., horns) which are controlled by single genes. The increase in variation between F_1 and F_2 for such quantitative characters as growth rate and milk yield will be very small, and in view of other sources of variation may not be detectable.

Certainly there is a danger that if the initial crossing is not done on a large enough scale then problems of inbreeding may be encountered. This is particularly the case if the breed is based on a single herd or on only one or two imported bulls. Inbreeding leads to decline in fertility, viability and growth rate to a greater or lesser extent. This has often not been realized and inbreeding has been deliberately employed in order to concentrate the blood of outstanding sires or to produce uniformity in a new breed. But uniformity is an elusive goal, a will-o'-the-wisp. It must not be sought at the expense of productivity. Indeed, it is essential at the outset to have as much variation as possible in order to allow scope for selection. The aim of crossbreeding is to combine the high yield of the European breed with the resistance of the local breed, and intense selection is needed to find the few animals combining both characters.

However, one should not be put off by takes about the formation of the Santa Gertrudis breed, which is said to have been based on a herd of 25 000 cattle on a ranch of 400 000 hectares (a million acres). In fact, only a few hundred of these animals were used for breed formation and the rest of the herd was merely graded up. The lesson from the Santa Gertrudis (as from the Shorthorn) is the danger of concentrating too much on single bulls. Both breeds were linebred to a single bull — *Monkey* for the Santa Gertrudis and *Shakespeare* for the Shorthorn — and it is probably no coincidence that both breeds have at times run into severe infertility problems.

If a breed is to be formed in a single experimental herd it is therefore essential to use a large number of bulls (at least 10-15) in the formative stages. As early as possible the programme should be extended to other herds and bulls should be interchanged between herds. A large cow population will be needed for progeny testing the bulls. Above all, animals should be selected not according to colour, conformation or other fancy points but for important economic characters which define total dairy merit, i.e., milk yield, viability and thrift of the calf, fertility of the cow, ease of milk let-down, udder and teat characteristics (including durability and suitability for machine milking, if appropriate), docility, and possibly also milk composition. Further details of a breeding scheme are available in the paper by Amble and Jain (1965).

At the present time crossbreeding between zebus and dairy breeds is taking place at several centres in the tropics. It would be most desirable to coordinate the activities within each crossing breed (Friesian, Jersey, Brown Swiss and Danish Red) and encourage exchange of breeding stock between countries to ensure a wide genetic base for each of the new breeds.

New Breeds by Introducing European Bulls Into a Zebu Population

There have been several attempts in the past to form new breeds by introducing European dairy genes into the zebu cattle of India and Africa, but none has come to fruition.

In Tanzania, for instance, Hutchison (Mahadevan and Hutchison, 1964) started a crossbred herd of European dairy cattle x east African zebus in 1946, but it was disbanded in 1966. Earlier the new strain had been officially baptized "Taurindicus". It is clearly easier to name a new breed than to maintain it, and the reports from India contain many attractive and descriptive names such as Jersind, Brown Sind and Karan Swiss which presumably will all be eventually absorbed in the Kamaduk. In tropical South America one has read of the Ocampo (Venezuela), Jerdi (Brazil), Tropical, Tropicana and Suisbu (Argentina). It would be interesting to have more particulars of these breeds and any others of European x zebu origin. While there are still no good examples of new dairy breeds whose genesis has been planned and executed on the basis of a European male X zebu female cross, there are two interesting examples of populations developed haphazardly from this cross in the past which have remained as distinct entities because of their economic value. These are the Taylor cattle of India and the Hatton cattle of Sri Lanka.

The Taylor cattle of Patna (Bihar state) originate from four Shorthorn and Channel Island bulls introduced by Commissioner Taylor in 1856 for crossing onto the local zebus. According to Sinha (1951) there were in his time about 2 500 Taylor cows, and their milk yield averaged 6–8 kg daily.

The Hatton (or Cape) cattle are said to derive originally from European imports which the Dutch brought to Ceylon via the Cape of Good Hope between 1765 and 1815. Until the more recent imports of European dairy breeds they were the principal milk producers on the island (Mahadevan, personal communication).

Neither of these populations has had the benefit of either a scientific breeding programme or a society for breed promotion. That they should have survived so long indicates that they must have considerable intrinsic merit. They deserve to be called "breeds" and merit a suitable improvement plan. Indeed, a programme of milk recording, progeny testing and breed promotion might even pay quicker dividends than starting now the laborious business of forming a new breed from scratch. Such a programme could, of course, include the introduction of outstanding halfbred bulls in order to benefit from the genes of today's improved dairy breeds.

New Breeds by Introducing Zebu Bulls Into a European Breed Population

The first cattle introduced into the New World (i.e., America and Australia) came from Europe. When it was observed that in the tropical and subtropical areas of these continents they performed less well than in their homelands it was thought that the explanation might be lack of heat tolerance. The remedy suggested was the introduction of zebu blood.

For this purpose the dairy zebus from Pakistan, the Red Sindhi and the Sahiwal, have been the most popular. Red Sindhis were imported into the southern states of the United States for experimental crossing with Jersey, Holstein-Friesian and Brown Swiss. The result of the cooperative experiments carried out in Georgia, Louisiana,

Maryland and Texas (Branton *et al.*, 1966) showed that as the proportion of zebu blood increased, the milk yield fell. It is true that heat tolerance increased, but this did not have any economic advantage. The differences in fertility, if significant, were in favour of the pure European breed.

These experiments have been given up, and the milk yield of European breeds in the North American subtropics is being increased by attention to feeding and management combined with selection for productivity within the hot environment.

Nevertheless, the conclusions of Branton *et al.* (1966) include the sentence: "If the hypothesis of lower nutritive requirements is acceptable, then under conditions of adverse climatic conditions and poor nutrition, some proportion of zebu breeding may prove desirable." There are two areas where this has proved to be the case and two new breeds, the Jamaica Hope and the Australian Milking Zebu (AMZ), have been developed.

At the government farm (Hope Agricultural Station) in Jamaica the beneficial effect on milk yield of using halfbred zebu bulls on Jersey cows was recognized before 1920, and in that year a single Sahiwal bull was imported to continue the good work (Lecky, 1949). Fortunately he turned out to be an outstanding animal, and the Jamaica Hope breed owes most of its zebu genes to him. The new breed contains about 20 percent zebu blood and 70–75 percente Jersey, the rest being from Friesian and other breeds. The bulls produced in the government herd were used to grade up the general Jersey-zebu stock on the island and now most of the dairy cows belong to this breed.

Mahadevan (1966) gives an excellent short account of the origin of this breed and Mahadevan *et al.* (1970) show that its further development is being limited by the restriction of progeny testing of bulls to the government herd. This means that the number of daughters per bull is too low for an accurate assessment, or else that too few bulls can be tested to given an adequate intensity of selection. All farmers should be involved in testing, by using young bulls extensively through artificial insemination and milk-recording their daughters.

In Australia the approach to forming a new breed based on Jersey and zebu has been much more systematic, and is described in the article by R.H. Hayman in this issue of *World Animal Review*. In fact, the first experimental results were similar to those in the United

States — in both cases the F_1 had a much lower milk yield than the Jersey. At this point the Americans gave up but the Australians were convinced that zebu blood had something to offer in the warmer, tick-infested parts of northern New South Wales and in Queensland. The results which Hayman gives, indicate that they were right. The breeding system used has been crossbreeding and selection up to the F_3 generation, followed by grading up the local population of Jerseys to the halfbred bulls and at the same time progeny testing these bulls. The question arises, why should zebu blood be an advantage in New South Wales (and in the West Indies) but not in the similar climate of the southern United States? Several differences can be noted which may be significant. For instance, cattle ticks and tick-borne fevers are present in northern New South Wales (and in Jamaica) but are not a problem in the United States. Feeding and management levels are clearly different in Australia and the United States, judging by the mean yield of 2 989 kg for Jerseys given by Branton *et al.* (1966) compared with 1 944 kg in Hayman's. In Australia there is more reliance on pasture. The F_1 cows also have more difficulty in establishing a lactation in the absence of the calf. In the United States the F_2; yielded less than the F_1; in Australia the F_2 and F_3 yielded more than the F_1. This increase was presumably achieved by intense selection among their sires.

New Breeds of Multiple Racial Origin

The above discussion of new breed formation is concerned with a foundation based primarily on two breeds only — one improved temperate dairy breed and one zebu. (The additional European blood in the Taylor, Hatton and Jamaica Hope breeds and the use of two zebu breeds in the formation of the AMZ, do not really affect the issue.) In a planned breed formation three or more breeds could be used as a matter of deliberate policy. This will increase the genetic variance and hence the scope for selection, but it will also make the programme more complicated and delay the final birth of the breed. It may also overload the foundation with genes from inferior breeds — note how Hayman had to get rid of the inferior Sindhi crosses.

The Indian Council of Agricultural Research has launched an ambitious programme to create a new breed — the All-India Coordinated Research Project on dairy cattle breeding (Bhat, 1972). It is based on the current crossbreeding between Jersey, Friesian and Brown Swiss on the one hand and various local zebu breeds on the other (Hariana, Sahiwal, Gir, Ongole, Hallikar, Tharparkar) at eight different centres. The first crosses will be backcrossed to a second European breed and the backcrosses combined in pairs. Thus it will be three generations (say 15 years) before the foundation stock is ready. Then will start the important phase of progeny testing bulls on a large enough scale so that those rare animals can be picked out which combine (in some measure at least) the genes for high milk and butterfat yield and growth rate and fertility of the European breeds with those for disease resistance and heat tolerance of the Indian breeds.

This is a very ambitious programme. It might be quicker and easier to think in terms of two new breeds — one based on the Jersey for early maturity and butterfat, the other on the Friesian for milk and beef. The preliminary results from the crossbreeding work at Haringhata, West Bengal, which was started under the auspices of an FAO project and has now been incorporated into the All-India Coordinated Research Project. If these are confirmed by later results they would appear to indicate the advisability of concentrating on Jersey and Friesian rather than Brown Swiss.

Crisscrossing Between Local and Improved Breeds

The examples given above show that while it is easy to increase the milk yield of a zebu breed by crossing with a European dairy breed the difficulty is to maintain the advantage in later generations. It needs an extensive system of milk recording and progeny testing, preferably based on an artificial insemination service so that each bull can have many daughters distributed through several herds. If no such facilities are available it is better to use a crisscross system of breeding.

When equilibrium is reached, the cows in alternate generations have one-third improved or one-third local blood. These two types should be kept in separate herds, the first run with a European bull and the second with a local bull. If this is not possible, the presence or absence of hump (assuming that the local breed is a zebu) will indicate which breed of bull should be used on each cow. If artificial insemination is used the separation into herds is not necessary; the humped cows will be inseminated with European semen and the humpless with zebu.

The total resources of the farm can be used for keeping cows and breeding their replacements. All male calves will be slaughtered and bulls will be brought in as necessary. The continuing genetic improvement stems from these bulls. They will come from herds in which a rigid breedimprovement system is in operation. If artificial insemination is available then semen can also be obtained from outside the country, but if the exotic semen comes from a temperate country the question of genotype x environment interaction must be borne in mind — the best bulls as tested under temperate conditions may not be the best in the tropics.

Naturally a source of improved local bulls is also necessary, and this highlights the importance of a selection programme such as that described by Meyn and Wilkins for the Kenya Sahiwal in this issue. Replacement cows will be chosen from the best cows of the two types. The two-thirds exotic may well give more milk than the one-third exotic, but the temptation to choose more two-third than one-third exotics as replacements should be avoided. This would mean a preponderance of onethird exotics in the subsequent generation. It would therefore be desirable to start the crisscrossing by backcrossing half the F_1 cows to the local and half to the exotic bull. This would enable the two types to be kept in equilibrium from the start.

Crisscrossing has very definite genetic advantages. By breeding always from crossbred cows it exploits any hybrid vigour in fertility as well as in yield. It can be modified by introducing a third breed into the rotation, e.g., a second European breed, if it appears desirable to maintain a higher proportion of European blood. Some results of the beginning of a rotational crossbreeding scheme using three breeds at Turrialba, Costa Rica. However, since the last cross was by the breed with the highest milk yield it is not clear how much of the advantage of the three-breed cross is due to additive gene action and how much to heterotic effects. A comparable system has been in operation among beef-cattle breeders in Africa for many years on an *ad hoc* basis. The defensive, almost guilty way they describe it is entirely unnecessary. The system is genetically sound and should be exploited for both beef and dairy cattle.

In their article in this issue, Meyn and Wilkins present evidence that in the semiarid and coastal areas of Kenya the Sahiwal-Ayrshire cross gives more milk than the purebred Ayrshire, and the Sahiwal-Friesian cross more calves than the purebred Friesian. They also demonstrate that Kenya has a source of improved Sahiwals and of locally adapted European dairy breeds. They therefore recommend a crisscrossing programme of the type outlined above, particularly for use in dairy ranching.

Likewise Mahadevan (1970) presents evidence of the advantages of a rotational crossbreeding system for the lowlands of Sri Lanka.

Such a scheme has the further advantage of flexibility. The breeds used for crossing can be changed immediately according to changes in environment or demand (milk or meat). The two types of animals produced may have different roles to play, e.g., the two-thirds zebu steer may be more suitable for draught, or in a dairy ranching enterprise the two-thirds zebu cows may be used for suckling.

If beef production is also important, then the crisscross cows not used for breeding replacements can be put to a "terminal" beef sire and all their progeny raised for beef.

The tradition of the uniform purebred should not be allowed to obscure the many advantages of this relatively novel method of breeding dairy cattle for the tropics.

6

Crossbreeding for Milk Production

The indigenous cattle of Sri Lanka, the Sinhala, in common with most other cattle of the tropics, are poor dairy animals. They are small, with an average adult weight of 160 kg. Their mean 305-day lactation milk yields are around 450 kg. This level of production, while it may satisfy the needs of subsistence farming, is too low for commercial dairying. It is therefore necessary to improve the production potential of the cattle in order to make them profitable.

The climatic conditions under which cattle are raised in Sri Lanka vary widely. The agroclimatic zones may be broadly classified into: dry zone, coconut triangle, midcountry and hill country. The ecology of these areas. About two thirds of the cattle are in the dry zone, 25 percent in the coconut triangle, and the remaining 8 to 10 percent in the midcountry and hill country areas where dairy managerial skills are well developed and high levels of concentrate feeding are adopted under systems of intensive husbandry. The coconut triangle, although hot and humid, is also regarded as an area of high dairy potential, partly because dairy farming has been a traditional enterprise of the farmers in this zone and partly because of the success in cultivating pasture grasses such as *Brachiaria miliformis* and *Brachiaria brizantha* as inter-crops. The coconut palms also provide a certain amount of shade and offer some measure of protection from solar radiation. In the dry zone the cattle forage on poor natural pastures with virtually no inputs of concentrate feed; managerial skills are limited, and milk is regarded as a by-product of animals that are raised mainly for beef.

The Government has large cattle farms of European breeds—primarily the Ayrshire, Jersey and Friesian — in the hill country areas that have a high dairy potential. The success that has been achieved during the last two to three decades with purebred herds of these breeds and with the upgrading of local cattle shows that high levels of European blood are not inimical to high milk production in this zone. The wellknown "Hatton" or "Cape" cattle found in the hill country are believed to be descendants of European cattle brought to Sri Lanka by the early Dutch settlers some 300 years ago. They acquired a certain amount of indigenous blood by crossing with the local cattle and were developed without any proper breeding policy or plan. The dairy merit of these animals lends support to the present policy of upgrading all the indigenous cattle in the hill country to European breeds.

V. Buvanendran is Animal Geneticist at the Veterinary Research Institute, Peradeniya, Sri Lanka; P. MAHADEVAN is Animal Production Officer (Research and Education), Animal Production and Health Division, FAO, Rome, Italy.

In the coconut triangle and the dry zone, climatic conditions are unfavourable for the adoption of a straightforward upgrading programme. The dry zone has the further disadvantage of poor nutrition, because forage in terms of both quantity and quality is limited mainly to the rainy season. Crossbreeding research has therefore been directed at evolving breeding policies that would be suitable for these two zones. The present article discusses the results of research carried out over the last 20 years and their possible application to national programmes. The experiments were carried out at the government livestock stations at Karagoda-Uyangoda, Undugoda and Wirawila.

Karagoda-Uyangoda Experiment

Karagoda-Uyangoda is located in southern Sri Lanka, at an elevation of 30 metres above sea level with a mean annual temperature of 26.7°C and a relative humidity of 78 percent. Annual rainfall averages 2 500 mm. Crossbreeding work at the government livestock station was initiated by Mahadevan (1953). Sinhala cattle were crossed with either Jersey or Friesian sires. The F_1 generation was bred inter se to produce F_2 generations of both breed crosses. In the Jersey X Sinhala crossbreeding programme, part of the F_1 population was backcrossed to the Jersey to produce a B_1 generation with 75 percent Jersey blood. In the Friesian programme, the F_2 animals were carried through to the third filial generation by inter-se mating among the F_2. The results obtained during the periods 1956-66 (Wijeratne, 1970) and 1968-73 (Buvanendran, 1975, unpublished) are summarized separately. It will be observed that F_1 animals from both the Jersey x Sinhala and the Friesian X Sinhala matings showed a remarkable increase in milk yield over the indigenous Sinhala cattle, but the F_2 animals exhibited a marked decline in yield from F_1 levels. Further interbreeding to the F_3 stage in the Friesian programme did not produce any significant change in performance from the F_2. Comparison of F_1 performance in the Jersey programme with contemporary F_2 and B_1 production levels respectively suggests that although backcrossing to the European parental breed gave relatively better results than the inter-se mating of F_1 animals, the levels of production attained in the B_1 were nevertheless considerably lower than in the F_1.

These results suggest that hybrid vigour is responsible for part of the higher yields in the F_1 and for the subsequent depression in the F_2 and B_1. The F_2 and B_1 animals lose 50 percent of the hybrid vigour present in the F_1, but the relative superiority of the B_1, animals over the F_2 may be attributed to the additional amount of Jersey blood. The data on calving intervals suggest that the additional Jersey inheritance in the B_1 did not lower reproductive potential.

Undugoda Experiment

This farm is located in the midcountry zone at 180 metres above sea level. The average annual rainfall is 2 500 mm. Mean monthly temperatures vary from 23.0 to 31.1°C and mean relative humidity is 71 percent. The cattle are housed in sheds and zero grazed on *Brachiaria brizantha* pastures. Red Sindhi cows were mated to Jersey sires to produce F_1 animals. F_2, B_1 and 5/8 Jersey offspring were also produced by mating F_1 animals inter se, backcrossing to Jersey or mating F_1 to B_1 animals. A purebred herd of Ayrshire was maintained contemporaneously and this served to provide a control population during the major part of the experimental period.

Data on milk yield and other characteristics of economic importance are present (Buvanendran, 1974).

The results are essentially similar to those of the Karagoda-Uyangoda experiment. The yields of the B_1 animals were lower than those of the F_1 but closer to the latter than to the F_2. The difference between the Karagoda-Uyangoda and Undugoda experiments is probably due to the favourable climatic and managerial conditions at Undugoda that permitted a fuller expression of the Jersey genotype in these animals. It may also be noted that the performance of the Jersey F_1 in this experiment is similar to that of the purebred Ayrshire.

Reproductive performance as measured by the length of the calving interval varied with the different crosses, but there was a general tendency for calving intervals to increase with an increase in *Bos taurus* blood.

The B_1 and F_2 animals in this experiment were born and commenced lactation at about the same time, so that the treatments accorded to them were similar. The differences in performance may therefore be regarded as largely genetic.

Wirawila Experiment

This government farm is located in the typical dry zone of Sri Lanka. Mean annual temperature is 27.1°C and average precipitation 1 075 mm. Breeding work at this station commenced with Sinhala animals, but these were first mated to Red Sindhi sires to produce females with varying levels of Red Sindhi blood (ranging from 1/2 to 7/8), and these were then crossed to either Shorthorn or Jersey sires.

The production characteristics of Red Sindhi X Sinhala crosses obtained during the period 1961–69, and of Shorthorn and Jersey crosses with the grade Sindhis during 1970–73, are summarize (Buvanendran and Tilakaratne, 1974, unpublished data). The contemporary records of Sinhala cattle are not available, because these animals had been disposed of at the time that the F_1 and higher grades had come into milk. Nevertheless, the results are indicative of the increase in yield obtainable by crossing the Sinhala with the Red Sindhi.

The yields of the F_1 are close to the expected mean of the two parental breeds (if it is assumed that the yield of the Sinhala parents at Wirawila would have been similar to that obtained at Karagoda-Uyangoda during the same period). This suggests a simple additive genetic effect on milk yields when two zebu breeds are crossed. Any heterotic effect is probably of a small order. Further upgrading to the Red Sindhi did not result in any further increase in yield. However, the introduction of European blood resulted in a marked improvement in milk production, the crossbred animals giving nearly 50 percent higher yields than the Red Sindhi. The reproductive performance of the European crosses was also superior to that of the Red Sindhi.

It is relevant to note that the performance of the Jersey crosses at Wirawila is higher than that of the Jersey crosses in the Karagoda-Uyangoda experiment during the same period, even though climatic and managerial conditions were less favourable at Wirawila. This is probably due to the superiority of the foundation zebu stock (Sindhi × Sinhala crosses) used at Wirawila. The Undugoda results also lend support to this interpretation.

General conclusions

On the basis of the results of the three experiments outlined above, the following general conclusions may be drawn:

1. Milk production from F_1 European x zebu crosses is markedly superior to that obtained from the base zebu population. Progeny from Friesian matings to the zebu produce more milk than those from corresponding Jersey matings.

2. Interbreeding of the F_1 progeny results in a marked decline in milk yield, the reduction in yield at the F_2 level averaging 30-40 percent. Further interbreeding to the F_3 generation does not lead to any further significant decline. These results point to an appreciable degree of hybrid vigour in the F_1 and its loss in the F_2 and F_3 generations.

3. Backcrossing the F_1 to the European breed results in performance levels that vary considerably with climatic and managerial conditions. Under relatively mild climatic conditions and with minimal exposure to heat stress (as at Undugoda under zero grazing), B_1 animals can produce at a level fairly close to the F_1, whereas under harsher climatic conditions (as at Karagoda-Uyangoda), B_1 production would be lower than F_1, although relatively higher than F_2.

4. Inasmuch as the genetic merit of the European parent has an influence on the performance of the F_1, the genetic merit of the zebu is also important, as evidenced by the results from Undugoda and Wirawila, where purebred Sindhis or Sindhi × Sinhala crosses were used as the foundation zebu stock.

National Breeding Programmes

In deciding on a national breeding programme to improve the indigenous zebu stock by the introduction of European blood, two options are possible — either new breed formation or the adoption of a crisscrossing or rotational crossbreeding system. Where new breed formation is desired, it is necessary to decide early on in the programme the European breed that is to be used for crossing with the indigenous zebus, and the level of European blood that should desirably be incorporated into the new breed. Where the evidence suggests that a system of rotational crossbreeding using two or more breeds is likely to be more fruitful than new breed formation, the decision concerns the number and choice of breeds to be used in the rotation.

For new breed formation, the choice of European breed should be made with the aim of maximizing yield in the F_1, so that despite the loss of heterosis in the F_2 and B_1 generations high yields would still be possible when the level of European blood that should desirably be incorporated into the new breed has been reached and inter-se mating takes place. From the experimental results obtained in Sri Lanka, it would appear that the Friesian would be more suitable than the Jersey as the European parent for areas, such as the coconut triangle, which have a high dairy potential notwithstanding the tropical climate. Prevailing meat prices could also influence the choice of breed.

For the dry zone, however, the Jersey has distinct advantages because of environmental and managerial considerations. It would also seem that using an improved zebu such as a grade Sahiwal or Sindhi as the foundation stock would pay dividends in the long run by ensuring higher performance in the F_1 and subsequent generations. This may be regarded as impractical on the grounds that it would delay the introduction of European blood into the programme by one generation. Nevertheless there are areas, such as the dry zone of Sri Lanka, where an immediate introduction of European blood is not feasible due to the prevailing poor standards of nutrition and management. In such areas the slower, two-stage process of improvement would be more suitable.

A similar conclusion was reached by Mahadevan *et al.* (1962) in their genetic study of the Sahiwal grading-up scheme in Kenya: they stated that the Sahiwal breed in Kenya provided a satisfactory first stage for grading up the East African zebu to a more productive level in areas of medium agricultural potential: the second stage would then consist of one or perhaps two introductions of European blood into the Sahiwal grade stock, depending on the level of management attained by the stock owners at the end of the first stage and on the general climatic conditions of the area. The decision on the level of European blood that should desirably be incorporated in any national crossbreeding programme in Sri Lanka would be governed by similar considerations. The lessons learned from new breed development in Jamaica (Wellington and Mahadevan, in this issue), concerning the need to ensure the participation of a large number of milkrecorded herds and their full involvement in sire testing, are particularly relevant to Sri Lanka where both milk recording and sire evaluation are largely confined to governmentowned herds. Jamaican experience has demonstrated that a relatively small nucleus of animals of a new breed, developed within one or perhaps a few government herds, is unlikely to make any significant impacton milk production in a time span of 20 to 25 years without the cooperative efforts of a number of participating farmers.

Rotational Crossbreeding

The second option for a national breeding programme for the coconut triangle and the dry zone of Sri Lanka — crisscrossing or rotational crossbreeding, using zebu and European breeds — appears to be better according to the experimental evidence. This conclusion is based on two principal considerations. The first is the large decline in performance from the F_1 to the F_2 generation in both the, KaragodaUyangoda and Undugoda experiments, which indicates the presence of a high degree of hybrid vigour. The second is the relative superiority of the B_1 animals over the F_2 although the magnitude of this superiority varies with climatic and managerial conditions.

Two practical questions would need to be answered in designing and implementing a rotational crossbreeding programme, namely, the number of breeds that should be used in the rotation and the choice of breeds. A two-breed rotation involving a zebu and a European breed has the advantage that it is simple to operate, and when equilibrium is reached it should be possible to distinguish the cows with two thirds European and one third zebu genes from those with one third European and two thirds zebu genes; this would facilitate the choice of sires that should be used on each type of cow. Furthermore, there is no experimental evidence to suggest that three-breed combinations will offer greater advantages than twobreed systems. But there is one major disadvantage in a two-breed rotation using widely different parental strains. This arises from the fact that there would be two kinds of animals in the population at all times, and since their production levels are likely to be substantially different it may be psychologically and commercially difficult for producers to continue with a two-breed rotational crossbreeding system. It was partly in this context that a three-breed system involving one zebu and two European breeds was recommended by Mahadevan (1970). For a two-breed system, the breeds most suitable for the coconut triangle of Sri Lanka would be the Friesian and the Sinhala, with the proviso that if the Sinhala could be upgraded to an improved zebu before being incorporated in the rotational crossbreeding programme substantially higher yields are likely in the long run. Under dry zone conditions, the choice of the Jersey as the European breed component of the rotation has much to commend it because of the harsher environmental conditions. For the three-breed system, an improved zebu such as the Sahiwal × Sinhala or Sindhi × Sinhala cross may be used, together with the Friesian and the Jersey.

Whatever rotational crossbreeding programme is finally adopted, it would have the advantage that the whole of the national herd in a given zone could become involved in the programme even in the absence of a milk recording system. The role of the government livestock stations would then be to raise nucleus herds providing superior sires of the chosen breeds for crossing, rather than serve as centres for the multiplication of breeding stock for distribution to farmers. The importation of improved sires or their semen from other countries would serve the purpose of maintaining the quality of the nucleus herds and increasing their contribution to the national breeding programme.

The Development of the Australian Milking Zebu

Tropical climates are generally inimical to dairy cattle originating from temperate zones, while cattle native to the tropics, although genetically well adapted to the stresses of climate and parasites which they encounter, are not high milk producers. Australia, which has approximately one third of its land mass lying north of the tropic of Capricorn, has no indigenous cattle and, as would be expected in a country settled by Europeans, the cattle introduced were Bos taurus.

As dairy animals, these did not perform as well as expected in the country's tropic and near-tropic areas and it was decided to make an attempt to improve their performance by introducing exotic (Bos indicus) genetic material, and by crossbreeding combined with selection among the filial generations to establish a new breed which would combine the hardiness and resistance to parasites of Bos indicus with the higher milking potential of Bos taurus. Two breeds, the Red Sindhi and the Sahiwal, small herds of which had been presented to the Commonwealth of Australia by the Government of Pakistan, were used as Bos indicus parent material. Both are accepted dairy breeds on the Indian subcontinent. Jerseys were chosen as the Bos taurus parent. They were the most common dairy breed in Australia, particularly in areas of climatic stress, and were also generally accepted as being best able to withstand such stresses.

First Stage

In the first stage, which was carried out on the F.D. McMaster Field Station of the Commonwealth Scientific and Industrial Research Organization (CSIRO) at Badgery's Creek, New South Wales, parental breeds were crossed and the filial generations mated *inter se* up to F_3, Sindhi and Sahiwal crossbreds being kept separate. All females in each generation were milked for at least their first lactation, and males for breeding were selected on the basis of their dams' performance. Milk production under Australian dairy conditions was the sole criterion of excellence.

This stage, which has been described elsewhere in detail (Hayman, 1972), revealed that Sahiwal crosses were on the whole superior milk producers. The breeding programme has in later years been concentrated on animals derived from Sahiwals. Although the better Sindhi crosses have been retained and their offspring used in the general breeding programme, crossbreeding with Sindhis has been otherwise discontinued. In both crosses there were considerable differences in production between the daughter groups of the various sires used, indicating that selection for milk production would be successful. A major obstacle to progress was encountered when it was discovered that a high proportion of the crossbred females from both breeds failed to continue in lactation after separation from their calves. This appears to have been due to their failure to let down milk without the stimulus provided by the presence of the calf, rather than any inherent lack of producing capacity (Hayman, 1973). This characteristic is not a handicap in regions where Bos indicus is native, but is a complete disqualification under Australian dairy conditions where cows are permanently separated from their calves six to ten days after the latter are born. It was exhibited by some 70 percent of filial generation females, all of which were culled from the herd.

Altogether 212 Jersey females were mated to 10 Sahiwal and 8 Red Sindhi sires, to produce 146 F_1, F_2 and F_3 Sahiwal cross female descendants, and 124 F_1, F_2 and F_3 Red Sindhi cross female descendants. Average first lactation production figures for these animals. Data from all lactations, including the 70 percent of filial generation animals which failed in their lactations, have been used to compile the table.

Selection Differentials

Production figures from the animals retained in the herd were good, the best crossbred (an F_2 Sahiwal) having produced 4 649 kg of milk and 242 kg of butterfat in 305 days during her first lactation, which compares with a first lactation of 4 536 kg of milk and 242 kg of butterfat in 305 days from the best Jersey. Prospective sires were chosen from the male progeny of the best of these females. Selection differentials were not as great during the early years of this stage as they were later, when larger numbers of suitable females became available. However, maximum selection pressure was exerted at all times and numerous sires (8 F_1 and 6 F_2 Sahiwal, and 7 F_1 and 3 F_2 Red Sindhi crosses, respectively) were used to obtain as wide a sample as possible of the available genetic material. The selection differentials have been estimated from production figures which include the first lactation records of all dams.

Data have been used from animals which failed to let down milk after being separated from their calves. Thus, selection is against poor dairy temperament as well as low production, and the two have not been separated. This selection differential is not comparable with that found in standard European herds, but there is no satisfactory way of distinguishing the two classes of animal. On the other hand, the selection differentials are reduced on the male side when females whose production would qualify them as bull breeders produce only female offspring. Their records consequently appear in the general mean instead of appearing as dams of sires. At this stage, progeny testing of prospective sires was out of the question: young bulls were required for mating with heifers of their own age and generation, and it was physically impossible to obtain progeny test data for this purpose.

Second Stage

he second stage of the breeding programme was based on the progeny testing of young bulls from meritorious dams. In the later years of Stage 1 more young bulls of prospective merit were being born than could be used on the Field Station and arragements were made, with assistance from the New South Wales Department of Agriculture, to progeny test them in the herds of a group of cooperating dairy farmers in the Lismore area of northern New South Wales. Ten farmers, whose herds now number approximately 700 head, are in the group.

The objective in Stage 2 was to progeny test six bulls per year. As selected proven sires became available from the test, the best in each year was mated with the top 10 percent of producers in each of the cooperating herds in order to produce young sires for future progeny testing. The Badgery's Creek herd was maintained as a high-producing nucleus and source of young bulls with a higher level of Sahiwal blood (50 percent) than that available from the Lismore herds.

Two Screening Tests

At the height of this stage of the programme up to 40 young bulls a year were becoming available for consideration as progeny test sires, and it was possible to extend the criteria of selection by exposing them to two screening tests, one to artificial climatic stress and the other to infestation with cattle tick.

Bulls were exposed to climatic stress in groups of four. Each group was first placed in a room at controlled dry and wet bulb temperatures of 21 and 18°C, respectively, for a week, and the following parameters were measured for each individual: rectal temperature, sweating rate, respiration rate, and daily food and water intake. The group was then transferred for a second week to a similar room in which dry and wet bulb temperatures were 38 and 24°C, respectively. The same parameters were again measured, and the animals ranked in order of least departure from the first week's parameter. The two last animals on the list were discarded and the remainder retained for further testing under higher levels of stress. At the conclusion of the testing period 12 animals were retained and these were tested for tick resistance. The development and implementation of this testing procedure has been reported in detail by Allen and Donegan (1974).

This second screening for tick resistance, which has been shown by Hewetson and Nolan (1968) and Hewetson (1972) to be heritable, is carried out at the New South Wales Department of Agriculture's Cattle Tick Research Station at Wollongbar, in the area where the bulls are to be progeny tested. A critical requirement for the test is that the animals be maintained at pasture, being yarded for infestation and for counting of engorged females. The use of individual moated stalls, although desirable in certain specific experiments, may produce a degree of emotional stress which appears to influence the number of engorged females dropped by particular animals. For this reason screening tests are carried out wholly in the field. The bulls surviving the first screening are artificially infested with tick larvae, and then subsequently challenged with further infestations. The animals are ranked for resistance in inverse order of the number of engorged female ticks dropped at the challenge infestations. The seven highest ranking bulls are retained for entry into the progeny test. Of these the least fertile bull is dropped after the first nonreturn to service period. Progeny test matings commenced in 1962 and the resulting heifers had their first lactations in 1965, lactations in this stage of the programme being completed in 1972. The number of bulls in progeny test matings departed from the planned number of six per year, due to a variety of causes but mainly because in the early test years there were not always six bulls available. The actual number of sires tested each year and their highest and lowest contemporary comparisons, which also includes data from daughters of the first sons of selected sires (1972/73).

Production Data

A detailed study has been made of the production data obtained during the course of Stage 2 of the breeding programme. This has been published (Franklin et al., 1974) and shows that heritabilities for milk and total fat production are of the order of 0.23–0.27. These values must be considered satisfactory from the point of view of developing a new breed. A brief picture of the results obtained during this period is presented below. Between 1965 and 1972 first lactations were recorded from the daughters of progeny test sires. These were all obtained under the normal dairying conditions of the north coast of New South Wales, where yields are low by generally accepted standards. The low mean yields (1 375–1 820 kg milk) are a reflection of the practice of open grazing under hard conditions, where production and rainfall vary greatly from season to season. No pressure was brought to bear on cooperators to replace their Bos taurus cattle with the Australian Milking Zebu (AMZ), but virtually all have done so now. This argues, perhaps even more than comparative production figures, that this animal is meeting requirements.

The nature of the progeny testing programme did not allow worthwhile comparisons between AMZ heifers and Jersey, although it is recognized that such comparisons would have brought out clearly the relative merits of the two groups. But some comparative data are available from Badgery's Creek. First lactation production figures for all Jerseys milked on the F.D. McMaster Field Station, together with records from AMZ bred and reared there, and comparable groups of Jersey and AMZ heifers bred in the Lismore area, but reared and milked on the Field Station. That there is little difference between Jersey and AMZ heifers under conditions of optimum management. In other words, the Bos taurus potential for production is equalled by that of the new breed. Means for all records, including short lactations, are shown in the table and demonstrate the effectiveness of selection against milk let-down problems. In F_1 F_2 and F_3 females some 70 percent of the population were discarded because of short lactations. This figure fell to less than 10 percent in AMZ heifers, and sires currently selected from the progeny test include some whose daughters have all failed to exhibit the trait.

Third Stage

The third stage of the breeding programme is now in progress. In this stage only the sons of sires selected in the progeny test, out of females selected for their own high production, are being submitted to screening and progeny testing.

The final object of the exercise is the development of a dairy animal containing between three eighths and one half Bos indicus blood, and selected for milk production, tolerance to hot climate stress, and resistance to ticks. The cooperating herds now form a nucleus from which females are being disseminated to other herds in northern New South Wales and Queensland. Some animals have also been sent to Malaysia. At the same time the selected sires are being placed in government AI centres. Semen from these sources is now being used both in Australia and overseas. Young bulls with satisfactory hot room and tick resistance performance, but surplus to progeny test requirements, are being placed in Australian dairy herds which have difficulty in gaining access to AI. Surplus females from cooperators' herds are also being made available to other breeders in northern New South Wales and Queensland, and the breed is gradually spreading.

Breeding for Milk with the Sahiwal Stud

Kenya's cattle population of over 9 million head is kept under extremely varying ecological and management conditions. The objectives of cattle husbandry are equally diverse. The ecological zones range from the Afro-alpine moorlands on Mount Kenya to the very arid areas in the north; they include the highly fertile humid and subhumid agricultural areas of the highlands, the semiarid and arid range areas, and the fertile coastal strip. Farming systems range from subsistence agriculture and pastoralism to large-scale commercial farming and ranching. The objectives of cattle husbandry vary gradually from the traditional emphasis on numbers to higher production levels per animal and to the generation of income. The rapid increase of the human population and rising income expectations call for a better utilization of the cattle resource.

Several factors contribute to the low efficiency of Kenya's cattle herds: disease, overstocking, limited knowhow and capital, and a low genetic potential. The artificial insemination service in the areas with high potential and a regular supply of breeding stock for natural service in the range areas are efficient delivery systems for improved genetic material. To overcome the other constraints a much greater extension effort is required, which in turn will also emphasize the need to breed higher producing cattle.

Ecology and Genotype

The question that must be asked is whether this is feasible and economical. For the foreseeable future, cattle owners in Kenya will have to think in terms of production systems in which cattle utilize the existing environment with relatively small capital inputs. Once a shortage of land begins, and when better technical knowledge has been accumulated, more intensive production systems can be considered.

The cattle breeding policy for each ecological zone and production system in Kenya has been formulated on the basis of research findings and some educated guesses about the requirements of the people and the potential of the environment to meet these, Permanent field recording is necessary to check whether the cattle breeding policy keeps in line with development trends.

High-Potential Areas

Over one third of Kenya's cattle are kept in the high-potential areas where small holders and large-scale farmers derive their main income from crops. Only productive cattle are competitive. The climate is suitable for European cattle, dairy breeds being favoured by the present milk/beef price ratio. The future outlook favours dual-purpose cattle, especially on large-scale farms. Rapid up-grading with Ayrshire. Friesian, Guernsey and Jersey from the indigenous zebu is taking place. However, results from a large wellmanaged farm in a high-potential coffee-growing area (1 100 mm average annual rainfall) indicate that Sahiwal/Jersey crosses outyield the Jersey. It is thought that purebred European cattle will outyield the Sahiwal crosses on a higher plane of nutrition, but surveys carried out on smallholdings in the high-potential areas indicate that the average yields per grade cow and year are in the order of 1 350 kg, i.e., well below the production level on this farm. Only a few top herds reach averages in excess of 4 000 kg milk per cow and year.

There are now about 800 000 socalled "grade" cattle in this zone which have more than 50 percent European blood. Less than 300 000 of these are kept on large-scale farms: the majority are kept by smallholders. Over 20 000 breeding females are currently being reached by the rapidly expanding AI service. Most of the AI bulls have to be imported. Currently, bulls originating from 11 countries and five continents are standing at the Central Artificial Insemination Station.

Despite the low production level, there is genetic variation. With improving management and feeding conditions this is expected to increase. For a number of reasons it has been found desirable to organize an AI breeding programme for the three leading dairy breeds, the Ayrshire, the Friesian and the Jersey. Production conditions are so different from those in developed countries that genotype-environment interactions cannot be ruled out. A progeny test trial to answer this question will commence soon. In the meantime, each dairy bull recruited by the Central Artificial Insemination Station undergoes a progeny test on government stations and parastatal farms. The aim is to get 30 heifer lactations for evaluation per test bull and to select one sire out of five tested for semen production. For the time being the test capacity of about 30 bulls per year and the rapid expansion of the AI service do not permit this selection intensity. Further expansion of the officially milk-recorded cow population from the present 7 000 head will be essential. For financial and genetic reasons it is desirable to recruit more bulls bred within the country for the AI service. Bull mothers are selected on performance and conformation, and are bred to imported semen from outstanding bulls of the breed. Bull calves born to contract matings are delivered to the Central Artificial Insemination Station at the age of four weeks after which they are reared under uniform conditions. Mild culling on growth performance is planned for later years. When the homebred bulls are ready for service, they enter the progeny test programme along with the imported bulls.

Semiarid and Coastal Areas

Most of these areas are occupied by subsistence farmers and pastoralists. Milk for home consumption and feeder steers for the market are the main products. Crops are too unreliable to provide a livelihood without livestock. Cattle crossbred between a European dual-purpose breed (Friesian, Simmental, Brown Swiss, Red Poll) and the Sahiwal are considered to make best use of the environment. While Meyn and Wilking (1973) reported that at the Ngong Livestock Improvement Centre in the semiarid Kenya highlands first lactation milk yields were 1 519, 2 152, 2 233 and 2 466 kg for Sahiwals, Ayrshires, Ayrshires/Sahiwals and Sahiwal/Ayrshires respectively, Wilkins (unpublished) found the production data on a well-managed farm in a semiarid area to be as set out.

It has been concluded that a dairy herd of Friesian cows would produce slightly more milk than a herd of crossbreds, but the additional number of calves weaned would favour the crossbred unit in its overall economy. It was also reported that the Sahiwal bulls used in the herd were far inferior to those presently used at the AI station, which have been selected through the breeding programme. In a different study, Kimenye (1973) found that in a well-managed herd on the Kenya coast lactation milk yields and calving intervals for Ayrshire, 3/4-Ayrshire 1/4-Sahiwal, and 1/2-Sahiwal 1/2-Ayrshire cows averaged 3 024, 3 151 and 3 004 kg, and 424, 405 and 404 days, respectively, thus favouring the 3/4-bred Ayrshires. The formation of a new breed from a crossbred foundation is not considered, because of changing management and economic conditions. By the time a long-term selection programme yields its first results, farmers may need types of cattle different from those selected. By crisscrossing a European breed and the Sahiwal, farmers will benefit from genetic progress made in the selection programmes for existing breeds in the high-and low-potential areas or abroad. Flexibility in choosing the breed generation by generation will permit stockmen with a high standard of management and nutrition to increase the proportion of genes from the European breed, while under inferior environmental conditions farmers may decide to increase the proportion of Sahiwal blood. Some large-scale operations are practising dairy ranching, i.e., milking their beef cows once a day. In the vicinity of a milk market this system is very economical. It is basically the same production system practised by the pastroalists, and the appropriate breeding policy would be crisscrossing a European dual-purpose breed with the Sahiwal.

Arid and Very Arid Areas

The arid areas are of low agricultural potential, mainly occupied by seminomadic and nomadic pastoralists whose diet consists of milk, meat and blood. The 1.9 million cattle in these areas are kept for subsistence, milk and feeder cattle production. European cattle and their crosses with the Sahiwal probably have nutritional requirements that are too high for them to exist and produce in this environment. The Sahiwal is thought to be a suitable dual-purpose dairy/beef zebu for the low-potential areas. Breeding stock is disseminated for natural service by the National Sahiwal Stud at Naivasha, totalling about 80 bulls per year, and for artificial insemination by the Central Artificial Insemination Station at Kabete, using only proven sires.

Environmental conditions are believed to be too harsh for any genetie improvement in respect of milk production among the 1 million indigenous zebus in the very arid areas. Natural selection for survival and fertility has probably produced the most efficient cattle population.

The National Sahiwal Stud

Zebu cattle with good dairy characteristics are desirable for purebreeding in the arid areas and for crossbreeding with European breeds in the semiarid areas. They may even be needed for crossbreeding in high-potential areas. Cattle for these purposes are being selected for milk and beef production at the National Sahiwal Stud at Naivasha. Attempts to develop an improved indigenous dairy zebu in a number of livestock improvement centres came to an unsuccessful end in 1939 when four Sahiwal bulls were imported by the Government from Pusa, India, for an upgrading programme. By 1963, 60 bulls and 12 females had been brought from four herds in India and Pakistan. Early successes in upgrading the indigenous zebu (Mahadevan *et al.*, 1962) were followed by disappointments. Despite careful pedigree selection, some bulls with an undesirable dairy temperament were brought in from all four places of origin. This resulted in a high percentage of their progeny failing to let down milk in the absence of their calves.

The formation of the National Sahiwakl Stud at Naivasha in 1962 concentrated the breeding work on one farm, and was the start of a scientific breeding programme for a dualpurpose zebu to be kept under extensive pasture conditions, but under good management. A progeny test programme for milk production commenced in 1965 following recommendations by Mason (1965). The current breeding programme has operated since 1968.

The main objectives of the stud are:
* to develop a management system for extensive dairying in a semiarid environment;

- to make genetic progress for milk and beef production:
- to supply progeny-tested bulls to the Central Artificial Insemination Station:
- to supply breeding stock for range development areas (e.g., Masailand).

Management

The stud consists of 500 cows plus followers, a total of about 1 300 head. All heifer calves born and about 80 bull calves are reared per year. At birth the calves are immediately removed from their dams, weighted, and their identification number is tattooed in their ear. They are weighed weekly until they reach 55 kg liveweight, and thereafter fortnightly until 125 kg liveweight. They are dehorned with a hot iron at six weeks of age.

Calves are weaned from whole milk on to concentrates at nine weeks of age, having consumed about 210 kg milk in the rearing period. Concentrates are offered in the fifth week of life and they are consuming about 0.5 kg per day at the time of weaning. This gradually increases to 1.4 kg per day, and they are fed at this level until they reach 125 kg liveweight. Calves are reared on natural pasture. Only during times of drought do they receive lucerne hay as a supplement.

The calves are dosed with phenothiazine every time their pasture is changed and finally when they reach 125 kg liveweight. They are then branded on the left leg with the same identification number they received at birth. Bulls and heifers are sent to separate farms where they graze indigenous pasture without supplementary feeding for their remaining rearing period. They have access to a balanced mineral lick. Bulls are selected at two years of age, either for progeny testing or for natural service in the range development areas. All heifers are inseminated at 27 months. Seasonal calving is not practised. Pregnant cows are drafted into the milking herd for one month and heifers for two months before calving. The longer period for the latter allows them to become fully accustomed to the milking surroundings. Small quantities of concentrates (0.5 kg) are on offer.

All cows and heifers are weighed immediately after calving. They are bred again 70 days after parturition and are dried off at 305 days of lactation. The herd is milked in the field twice a day through four mobile bails, each manned by six hand milkers and a recorder. Individual yields are recorded at every milking and butterfat analyses are carried out once a month. The indigenous pasture is expected to provide for maintenance and the production of 5 litres of milk per day throughout the year. After prolonged rain, concentrates are fed only to cows exceeding 9 kg milk per day, at the rate of 1 kg for 2.5 kg. milk produced.

The milking herd has the best pasture available. Good paddocking permits rotational grazing, with the cows rarely staying longer than seven days in one paddock. Dry herds follow the milking herd to eat the less palatable grasses.

Productivity and genetic parameters

The collection of production data over a number of years has permitted the estimation of genetic parameters. The averages, standard deviations and heritability of some production traits.

The existing variance and the corresponding heritability estimates give promise of a rapid genetic improvement in respect of milk yields, and to a lesser extent in respect of growth rates. This is also demonstrated by the increase in yields since the systematic selection programme commenced in 1968.

Breeding Plan

The breeding plan for milk and beef. About 210 heifers enter the milking herd every year. The higher yielding 110 heifers are allowed a second lactation, after which the best 55 cows over the first two lactations are retained in the herd. Thereafter, hardly any culling is done on yield. Heifers and first calvers are inseminated with semen from test bulls, while all cows calving for the second time or more — the so-called elite herd — are inseminated with semen of proven bulls to produce the next bull generation. The 180 elite cows produce 150 calves, i.e., 75 bull calves per year. Of these, about 70 survive the bucket-rearing period and range-performance test for growth rate to an age of two years. The best 15 are then identified by a selection index, comprising the estimated breeding values for milk yields of the sire and dam, and the estimated breeding value for weight for age of the young bull itself. An inspection of the selected bulls and their dams for physical faults, muscling, udder conformation and teat shape eliminates three more bulls. The remaining 12 are tested for semen quality, leaving about 10 for test mating. The aim is to produce 14 milking daughters per test bull.

The average daily gain to weaning, the bulling heifer weight at 27 months, the 30-day milk yield and the 305-day milk yield of the first lactation are analysed by contemporary comparison. First decisions on test bulls are made when the 30-day yields become available. Bulls with outstanding daughter performances are sent to the AI station for semen production, while all the remaining bulls are kept on. Final decisions are made on the results of the 305-day milk yields. Two bulls are selected annually for the AI station and for use in the elite herd. The remainder are sold for natural service or for slaughter. Genetic progress in terms of milk yield may be expected at 3 to 4 percent per year (Meyn, Were and Bartilol, unpublished).

In recent years the demand for Sahiwals from the range areas of Kenya and from abroad has risen sharply. Deep-frozen semen or breeding stock have been exported to many countries, mainly in tropical Africa. There are plans to extend the Sahiwal breeding programme to other government farms. This could lead to even faster genetic progress.

The national dairy cattle breeding policy for the different ecological zones of Kenya is based on findings which suggest that, with adequate management, best use of pasture is made by European dairy breeds in the temperate high-potential areas, by crosses between European breeds and the Sahiwal in the semiarid and coastal areas, by pure Sahiwals in the arid areas, and by indigenous zebus in the very arid areas. The Sahiwal has a very important role to play, both as a purebred and as a crossbred with European breeds.

With the existing variation and the corresponding genetic parameters, rapid progress may be expected in the breeding programme of the National Sahiwal Stud at Naivasha. The overall herd average has increased from 1 042 kg milk per cow and lactation in 1965 to 1 630 kg in 1972. A future rate of genetic gain of 3 to 4 percent per year has been projected. Sahiwal breeding stock and deepfrozen semen are being exported to many tropical countries.

Development of the Breed of Dairy Cattle

Apart from the Taylor breed of India, which was evolved a century ago as an interbreeding population of high producing crossbred cattle (Henderson, 1927), and the newly developed Australian Milking Zebu (Hayman, 1974), the only other tropical dairy breed that has been successfully produced from a *Bos taurus X Bos indicus* crossbred foundation is the Jamaica Hope. This breed, which is supported by a breed society, has some 6 000 registered females and is fairly widely distributed throughout Jamaica. The demand for Jamaica Hope animals from other tropical countries is growing. Some animals have been exported to countries within the Caribbean area as well as to Latin America, but the growing demand can only be met by the export of bulls and semen. This article is concerned with the origin and development of Jamaica Hope, its performance characteristics, the breeding policy currently adopted and the lessons to be learnt for new breed development elsewhere.

Origin

The first phase of development of the Jamaica Hope breed began in 1910 with the testing of various European cattle and their grades under Jamaican conditions. The Government of Jamaica established a dairy herd at its Hope Farm, near Kingston, with Ayrshire, Brown Swiss, Guernsey, Holstein Friesian, Jersey and Red Poll cattle as well as crosses of these to cattle of mixed type (i.e., animals containing creole and zebu blood). This was followed in 1920 by the introduction of Sahiwal genes into the population through one bull imported from Pusa, India. There was insufficient evidence at that time to determine which of the European breeds was likely to contribute most successfully to the dairy industry of Jamaica. However, it soon became evident that the six imported European breeds could not all be tested on the same farm, and those with the poorest performance were eliminated at an early stage. The Ayrshire and Brown Swiss showed little promise over the Holstein Friesian and Jersey and were therefore removed in 1928. Red Poll breeding for milk stopped in 1938, although these animals continued to be used for beef breeding and contributed to the development of the Jamaica Red breed of beef cattle.

The second phase commenced in 1943, when it was decided to discontinue Guernsey breeding; the average age at first calving among Guernseys was 42.1 months and the mean calving interval was 14.9 months. The gene pool of the animals at the Hope farm was broadened through importation of more Jersey bulls, and considerable attention was also paid to the improvement of husbandry practices and to the testing of sires on the basis of progeny performance. The sires in service in 1943 were four imported purebred Jerseys and five Jersey-Sahiwal crosses ranging from 3/4 Jersey 1/4 Sahiwal to 1/8 Jersey 7/8 Sahiwal. The grade Jersey females varied in their inheritance from 1/4 Jersey 3/4 Sahiwal to the equivalent of four topcrosses of Jersey (Lecky, 1962). The average age at first calving among grade Jerseys was 38.3 months and that of purebred Jerseys 35.8 months. With improvements in feeding and management the overall mean was reduced to 32 months. Calving intervals averaged 14.2 months. The highest lactation yields were produced by grade Jerseys with 5/8 to 7/8 Jersey inheritance.

The breeding of Holstein Friesians was terminated at the end of the second phase in early 1952. The reasons for this were as follows. Although the half-bred Holstein Friesians gave higher milk yields than the Jersey crossbreds, there was a significant decline in yield as the level of Holstein Friesian blood increased. The culling rate was highest among animals of Holstein Friesian breeding, and this increased with the proportion of Holstein Friesian inheritance. The mean age at first calving was 42.1 months, and calving intervals increased in length from a mean of 15.2 months for all animals to 17.2 for three-quarter breds. Milk production per acre was also regraded as being lower among grade Holstein Friesians than among grade Jerseys.

Elevation to Breed Status

In 1950, the cattle at Hope Farm were transferred to the Bodles Animal Research Station and thereafter bred as a closed herd. The commencement of the third phase, in 1952, was marked by an official ceremony at which all the animals at the Bodles Station and those of similar genotype on privately owned farms throughout Jamaica were elevated to breed status. The basic population then consisted of both purebred and grade Jerseys, as well as of selected animals from the Holstein Friesian breeding programme. The grade Jerseys were those that had acquired genes for adoptation to a tropical environment through the admixture of zebu blood during the period 1910 to 1952; their lower age at first calving, shorter calving intervals and smaller body size, as compared with other animals, were regarded as evidence that under Jamaican conditions they would be more economical producers of milk than the general run of Holstein Friesians and their derivatives. The few animals of Holstein Friesian breeding that were incorporated in the breed were those that were descendants of matings to the foundation family of Norbrook Jerseys.

The third development phase, which is represented by the period 1952–64, involved considerable attention to breed development within the nucleus herd at Bodles. Sires were selected on the basis of dam and progeny performance and no attempt was made at fractional breeding.

The breed stabilized at a level of about 80 percent inheritance from the Jersey, 15 percent from the Sahiwal and 5 percent from the

Holstein Friesian. Impressive production records were achieved by individual animals; notable among these were Sam's Nymbrook and Stalin's Victory Noris. The most outstanding cow was Stardust, which produced a lifetime record of over 70 000 kg milk in 12 lactations.

The period 1964–74 may be regarded as the fourth phase in the life of the breed. During this period, efforts to increase the population of Jamaica Hope cattle and the total amount of milk produced were not significantly successful. The breed recorded lower levels of production in the Bodles herd, but some breed society members who participate in the breeding and registration of Jamaica Hope cattle improved their levels of husbandry and achieved higher production than was possible at Bodles. Enthusiasm for the development of the breed was lacking and cattle of Holstein Friesian breeding were imported in an effort to meet the island's need for the fresh milk.

These importations likewise failed to make the impact envisaged and emphasized the need for improved husbandry, irrespective of breed.

Breed Performance

National milk production recods covering the period 1965–73, computed through a locally based Dairy Herd Improvement Scheme, are set out, the Jamaica Hope breed. Selected herds of the breed in Jamaica were also recorded through the Dairy Herd Improvement Association of the United States Department of Agriculture, and the records processed in North Carolina. For 12 herds so recorded, the average 305-day milk yield based on 2158 records was 2737 kg, with a mean butterfat yield of 130 kg and a butterfat content of 5 percent. These figures are rather higher than the national average.

Earlier studies (Wellington *et al.*, 1970) carried out at the Bodles herd and in six cooperating farmer herds provided the parameters set out. The Bodles herd averaged 3218 kg milk in 305 days as compared with 2755 kg for the farmer herds. A larger sample of herds belonging to 34 breed society members was also recorded over the period March 1967 to February 1968 to determine the effect of variations in husbandry practices on milk production. The herds were first classified as good, mediocre or poor, according to level of management, and then recorded. Their 305-day milk yields averaged 2905, 2000 and 1623 kg respectively. These results confirmed the view that the breed had a reasonable potential for milk production under tropical conditions, but that good husbandry practices were essential for this potential to be realized.

Studies of individual herds showed considerable variation in age at first calving and in the length of the calving interval. The mean age at first calving ranged from 33.1 to 42.4 months and the calving interval varied from 371 to 466 days. The relatively late age at the first calving highlighted the need to improve the nutrition of heifers, the control of internal parasites during early life and conception rate. Research currently in progress aims at breeding heifers at 15–18 months of age to facilitate a mean first calving of 27 months. It is not expected that this will cause a significant decline in yield in the first lactation, as evidenced by the relationship between age at calving and first lactation milk yield in the Bodles herd. Furthermore, with a mean birthweight of 24 kg for female calves and a demonstrated rate of liveweight gain of 0.45 kg per day to age at first service (McLaren, 1966), it should be possible for heifers to be successfully mated at a body weight of 225–250 kg for calving at 27 months, even if this requires two services per conception. The lowering of the mean calving interval to 400 days or less would call for concerted efforts to reduce the service period. Greater attention to the nutrition of animals in early lactation, mating at the first heat following 60 days after calving and improved heat detection are likely to pay the best dividends.

From a disease standpoint, the Jamaica Hope has the advantage that it was exposed to tick infestation right through its formative stage and that all animals of the breed, both in the Bodles herd and elsewhere, were raised in an environment where anaplasmosis and piroplasmosis occurred. As a result, it has developed a remarkable degree of tolerance to these diseases.

Breeding Policy

At the inception of the breed, two criteria of selection were established: production performance and fertility. To date, no conscious selection for colour or type has been practised. In recent years it has become evident that udder conformation and quality should be included in the overall criteria of selection.

Selection of female breeding stock on the basis of their production performance has been negligible among Jamaica Hope cattle since the time of breed formation (Roache *et al.*, 1970). This is understandable in a situation where breed development continuos to be based on a relatively small population, where involuntary wastage is encountered and where the need to increase numbers is crucial. Voluntary selection among females would tend to be minimal in these circumstances.

However, the selection of cows as dams of herd sires offers greater opportunities for the genetic improvement of the breed.

Thus, cows with a minimum production of 4 545 kg (10 000 lb) milk in a 305-day lactation are currently selected as dams of potential herd sires and these cows are usually mated to proven bulls. Selection between sires is then made on the basis of their own progeny tests.

Selection and Testing

For many years, the entire population of Jamaica Hope cattle depended on the Bodles herd for sire service as well as for the evaluation of sire performance. The restriction of progeny testing to Bodles was found to hamper the further development of the breed (Mahadevan *et al.*, 1970); the number of daughters per sire was often too low for an accurate assessment of its breeding value, and too few sires could be tested to give an adequate intensity of selection. It therefore became evident that all farmers should be involved in testing, by using young sires extensively through artificial insemination and milk-recording their daughters. Selection and testing of sires over a wider cross-section of the total population is now in progress, employing either natural service or artificial insemination and analysing the results using contemporary comparisons. The number of daughters per sire on test, however, continues to be small and the method is therefore used only as a screening device for culling the poorest of the young sires.

Although some 50 breed society members maintain Jamaica Hope cattle, the degree of enthusiasm required for the further development of the breed is lacking within the country.

The Future

Most of the commercial cattle enterprises in the island do not participate in breed development; they change their breeds all too often and tend to vacillate between dairy and beef operations. In recent years, a large number of Holstein Friesian cattle have been imported into Jamaica from North America. Unless Jamaica Hope breeders are prepared to meet the demand for more breeding stock and improve their levels of efficiency, there is every prospect that this breed, line the Taylor breed of India, might be of only historical interest in the decades to come.

It would be relevant in this context to investigate the feasibility of an exchange of breeding stock between the recently developed Australian Milking Zebu and the Jamaica Hope, which have been evolved from similar foundation stock. If this could be organized, it would help provide a wider genetic base for both breeds. Several important lessons may be learnt from Jamaican experience for new breed development elsewhere, if such development is desired. The first relates to the initial problem of deciding which breed or breeds should be used for crossing. If the aim is to obtain not merely the highest yield of milk but the highest "overall dairy merit" and the highest market return under the prevailing local conditions, a single experimental herd would normally be inadequate for the testing of several breeds. Secondly, even after the choice of breed has been made, it would be unwise to confine the work on new breed development to one herd, except perhaps in the formative stages. A large number of sires (at least 10-15) of the chosen breed would need to be tested annually, and herd size thus becomes a limiting factor once again. It is therefore essential that the programme should be extended as early as possible to other herds; a large cow population is required for the effective progeny testing of a large number of sires and for the intense selection for traits of economic importance that must be practised. Jamaican experience also highlights the fact that unless the herds participating in the programme are efficiently milk recorded and become involved in the testing, by using young sires extensively through artificial insementation, their inclusion will not contribute materially to breed development. Indeed, in course of time, they should themselves be able to offer sires for testing and in this way broaden the scope for selection. Continued improvement from new breeds requires further selection within these populations. Without the Cooperative efforts of a number of participating farmers, a relatively small nucleus of animals of a new breed developed within a single herd is unlikely to make any significant impact in a time span of 20 to 25 years.

European Breeds

In recent times considerable numbers of *Bos taurus* dairy cattle have been introduced into many countries in the Near East where the prevailing species is *Bos indicus*. Since the climate in this region differs markedly from that in the natural habitat of the European breeds, the question has frequently been raised as to whether climatic factors have been responsible for the often disappointing results obtained from *Bos taurus* cattle. Alternatively, are these poor results due to inadequate management and feeding and lack of knowledge of the requirements of exotic cattle

This article reviews the results of a three-and-a-half-year study (Ansell, 1974) conducted in the United Arab Emirates, where the summer climate is extremely inimical to dairy production. The study was made on British Friesian cattle, which are not noted for their heat tolerance. In view of the relative success achieved in maintaining high production under such harsh climatic conditions, this article has been written for the benefit of those who plan to introduce high-yielding European cattle into other parts of the Near East or to countries with similar climates. It describes some of the problems likely to be encountered and the steps that can be taken to alleviate the stresses imposed on the cattle.

Physiological Considerations

Body heat in cattle is largely the product of the exothermic physical and chemical reactions within the animal and the absorption of heat by radiation and conduction from the environment. The dissipation of body heat is achieved by radiation, conduction and evaporation of moisture. Other minor methods of acquiring and dissipating heat are the voidance of faces and urine, raising ingesta to body temperature, and endothermic reactions. But these can be ignored in the context of animal management, with one exception: the ingestion of cold water, which could be of some significance.

Evaporation of moisture, whether from the skin or lungs, is a major factor in heat dissipation, and it is unfortunate that cattle, in particular the European breeds, are less well endowed with sweat glands than heattolerant mammals such as man.

Evaporation is one of course dependent on the humidity of the air and the wind velocity over the evaporating surface. Although this is a controversial subject and involves complex physical concepts, it can be safely assumed that relative humidity readings as taken from a wet and dry bulb thermometer can be used as a guide to the evaporative capacity of the air. It is this major dependence on evaporation for heat dissipation which makes high relative humidities — in association with high temperatures—so critical in producing heat stress. Various formulae have been devised combining humidity and temperature to produce a "heat index". Perhaps the most promising of these is that of Bianca (1962): heat index = dry bulb °C × 0.35 + wet bulb °C × 0.65. But neither this nor other formulae have been fully confirmed experimentally, although they to provide a crude assessment of comparative climatic severity when the introduction of exotic stock to unfavourable climates is being considered.

Conduction and radiation are simpler concepts of heat dissipation than evaporation. They are dependent on environmental temperatures and on the surface area/body weight ratio of the animal, i.e. an animal with a large surface area in relation to its bulk (e.g. a bat) can dissipate heat by these means more easily than one with a low surface area to body weight ratio. It has often been assumed in this connexion that the large dewlap, pendulous sheath, large ears and hump of Asiatic cattle which contribute to their increased surface area also contribute to heat tolerance. However, this is not necessarily the case, since experimental surgical excision of these organs has shown that the heat tolerance of the animals so treated is unaffected. This observation may be of some consequence in that it would be useless to select smaller animals from within a breed for introduction into hot climates solely in the belief that their surface area/body weight ratio would result in increased heat tolerance. The effect of such selection, if any, might be insignificant in relation to that of other factors.

An important aspect of heat dissipation is the diurnal variation in dry bulb temperature. Cattle can build up heat within their bodies to a considerable degree during the day without adverse effects if the night temperatures are sufficiently low to allow for its dissipation. But high sustained body temperatures are destructive.

A detailed knowledge of the climate of the area in question is therefore essential for assessing the heat stress likely to be encountered by animals. Monthly means of maximum and minimum temperatures are of little use; diurnal variations of temperature and humidity and their duration are of much greater significance. The results obtained in the United Arab Emirates suggest that on grounds of climatic factors alone, there are likely to be very few areas in the Near East where European breeds cannot be maintained successfully.

The severity of the climatic conditions experienced in the United Arab Emirates; they are near the upper limit of tolerance of Friesian cows in milk. It should be noted that the temperatures were recorded in well-shaded sites; they should not be equated with temperatures recorded in the standard "Stevenson screen" type meteorological box which, during the hours of sunlight, gives much higher readings.

Effect of Diet

The results of climatic chamber experiments conducted at the University of Missouri in the United States, and elsewhere, suggested that depression of appetite is a significant factor in reducing the milk yields of animals subjected to high environmental temperatures. These results were not corroborated in the United Arab Emirates study, since it was possible to maintain the animals on a constant diet throughout winter and summer with a ration that maintained both yield and body condition. This diet consisted of: a maintenance ration of 20.5 kg freshly cut alfalfa, 32 kg compounded dairy nuts, 1.8 kg bran, 14 g mineral and vitamin supplement and 57 g dicalcium phosphate; and a production ration of 1.8 kg of compounded dairy nuts per 4.5 kg of milk.

No grazing was available to the animals, and salt was not included in the ration since the salinity of the drinking water was very high and salt was in any case refused when offered.

This diet proved very satisfactory and resulted in a Compton metabolic profile well within standard limits. It is possible that the severe depression of appetite reported in the climatic chamber studies (which was not encountered to any great degree in the study) may have been a psychosomatic effect on the animals resulting from the close confinement inherent in chamber experiments. However, it was observed that the animals in the study took considerably longer to consume the ration in the summer months when conditions became extreme.

Reproduction

The suppression of spermatogenesis and early foetal reabsorption (Stott and Williams, 1962) are well documented under high temperature regimes. These phenomena were amply demonstrated in the first year of the United Arab Emirates study in that no successful conceptions took place between the months of June and October, although libido and manifestations of heat were not seriously affected. Two points are worthy of note in this connexion: spermatogenesis can be impaired for as long as nine weeks after exposure to heat (Johnston, Naelapaa and Frye, 1963); and oestrous periods tend to be "quiet" and short, thus making heat detection difficult if AI is to be practised.

It was also found that calving in the hot season subjected the animals to unacceptable stress conditions. Gestation periods were shorter by approximately 10 to 14 days for summer calvings, birth weights of calves were significantly reduced, and calves born during the height of the summer were found to be as much as 56 days premature in some cases. The shortened gestation periods adversely affected the dry periods prior to parturition. A high proportion of primary inertia was also encountered in the herd studied, but its relation to season of calving could not be demonstrated. It does not necessarily follow that all of these manifestations will occur in more moderate temperatures, but the possibility nevertheless remains.

Thus seasonal calving and close supervision of parturition are essential under climatic conditions as extreme as those encountered in the United Arab Emirates. Under such conditions, services should be limited to the months of February, March and April if neither conception nor parturition are to occur during the summer months.

Calf Rearing

Great care must be taken during the first 10 days of life of the calf because its thermoregulatory mechanism appears to be defective during this period, and it may develop extreme hyperthermia and die if artificial cooling is not provided.

For premature calves, this critical phase extends over a longer period. But once the calves have passed it, they tolerate heat stress well and pose few problems. Of 94 calves born during the period of the current study, none died after the first five days of life from any cause, and no cases of calf scours occurred.

Milk Production

Research workers have used a number of methods to attempt to estimate variations in milk yield attributable to different factors. Each has its limitations, and the best that can be said of any of them is that they only indicate approximate trends. In the United Arab Emirates study the depression of milk yield attributable to the climate was estimated by reference to the standard lactation curve constructed on the basis of the formula devised in the United Kingdom by the Milk Marketing Board (1969/70). On this basis it was found that the depression of yield attributable to the hot season amounted to the surprisingly low figure of 2.7 percent of the theoretical 305-day lactation, but this was almost entirely offset by the hitherto unsuspected "post-stress rebound".

The average yield of the 19 heifers having a 305-day lactation period in the present study was 3 253 kg. This compares favourably with the yield of 3 842 kg quoted as the average yield of 859 British Friesians used by the Milk Marketing Board to compile the standard lactation curve. A surprising factor was that the early summer months showed no reduction in yield; sometimes the reverse occurred. But toward the end of summer "fatigue" seemed to set in, and most of the reduction occurred at that time.

Health and Disease

Problems were experienced with hoof necrosis attributable to the animals standing for long periods on sawdust moistened with urine and dung. This is of relevance to areas where bedding is in short supply and where. animals must be confined on hard standings for long periods. It appears that most bedding at temperatures approaching bacterial incubation levels can be destructive to hoof tissues. No statistically significant difference was detected between the incidence of mastitis in summer and winter; the level of incidence was within reasonable limits.

An important health hazard was the pathogenicity of *Theileria annulata*, which is endemic throughout most of the Near East. Whereas mortality from this cause is comparatively low among indigenous cattle, it is very high among *Bos taurus* breeds. Tick control must therefore be complete and provision must be made for dipping or power spraying; hand spraying is usually inadequate.

No other special disease problems attributable to the unfavourable climate were encountered during the three and a half years covered by the study. It may therefore be concluded that if normal prophylactic medicine is practised, the incidence of infectious diseases should pose no difficulty. Nevertheless, it would be highly desirable to monitor the health of the herd by taking rectal temperatures daily. Since high body temperatures are to be expected under conditions of heat stress, it is imperative that a continuous record be kept to differentiate between the high temperatures caused by climate and those due to pathological conditions. In addition, care should be exercised in taking thermometer readings when ambient temperatures are higher than body temperatures, so that the mercury level does not rise in the thermometer after its removal from the rectum. A cotton swab soaked in alcohol is useful under these conditions.

Signs of Heat Stress

When the upper limit of heat tolerance is approached, cattle show some unmistakable and often alarming stress symptoms:

1. *Open mouth breathing*. This is indicative of very advanced stress; the head is extended, the mouth held open, and profuse salivation takes place.

2. *High respiratory rates.* These may be in the 100 to 120 range, but surprisingly they can be tolerated daily for long periods without causing any apparent symptoms of respiratory alkalosis.

3. *Second phase breathing.* This is a phenomenon first recorded by Beakley and Findlay (1955). At rectal temperatures of approximately 41.8°C there is a sudden drop in respiratory rate with a concurrent increase in tidal volume. This phase is also a symptom of advanced stress and should not be allowed to continue.

4. *Body splashing.* Strenuous efforts may be made by animals to paw water back onto their bodies from water troughs or to beat the water with their heads.

5. *Refusal to lie down.* This is a common phenomenon. Usually all the animals in the herd will stand for long periods when they would normally be expected to lie down; ablation of the "elbows" is often associated with this stage.

6. *Huddling.* Contrary to expectation, cattle subjected to heat stress will huddle together and will disperse as soon as the temperature drops to a moderate level.

7. *High rectal temperatures.* Afternoon temperatures of 40.5°C are not uncommon and can be tolerated for long periods providing that nightly remissions take place. However, if the rectal temperature exceeds 41°C, steps should be taken to reduce it.

Most of the above symptoms are valuable clinical indications of the degree of stress being experienced by cattle, and could serve as a guide to the time when steps to alleviate the stress must be taken. It may be noted in this connexion that lactating and heavily pregnant animals have a lower degree of heat tolerance than non-lactating animals. One *Bos indicus* animal in the study demonstrated its superior heat tolerance by lying happily in the sun with a barely perceptible respiratory rate of 8 to 10, when respiratory rates among *Bos taurus* cattle were in the region of 120. However, its growth curve was markedly depressed, and it could best be described as being almost in a state of aestivation.

Amelioration of Climate

Although the foregoing may appear to be a formidable list of woes likely to be encountered by cattle under adverse climatic conditions, in practice most of them were encountered only when deliberate efforts were made to refrain from taking steps to mitigate climatic stress. When such steps were taken, the herd remained remarkably trouble-free and tolerated the conditions of the Persian Gulf littoral very well. The more important methods adopted are listed below.

Showering: A variety of methods for cooling dairy cattle were tried:

a. complete air conditioning, (usually very expensive);
b. desert coolers, which rely on a forced draught over a wetted surface;
c. the provision of cooled drinking water;
d. breathing hoods supplied with a cooled air supply;
e. high-speed fans.

However, probably the most costeffective method consists of providing the animals with an artificial sweating mechanism to equate them with the freely sweating, heat-tolerant mammals. This can be done in a number of ways: the easiest is to invert a revolving domestic garden sprinkler on to the ceiling of a concrete-floored cattle shed. It was found that the cattle then formed an outward-facing circle at the periphery of the spray, and that a few hours free access to this in the afternoons reduced otherwise elevated body temperatures to normal limits, even in the relatively high humidities of the Gulf. The temperature of the water is to all intents and purposes immaterial since the latent heat of evaporation is utilized. The effectiveness of this method is amply demonstrat, which shows the mean rectal temperature of 21 cows with and without access to showering. The advantage of this method is that energy is not expended in lowering the temperature of large masses of air (as in desert coolers or air conditioners) which may not come into contact with the animal. However, a reasonable supply of water is necessary, but no more than is usually required if dairying is to be practised. Similarly, the temperature of an individual animal may be reduced dramatically by soaking it with water. It is not necessary to play a continuous stream of water over the animal, although if one wetting of the coat is insufficient, the coat may be re-wetted when dry.

Shelter engineering: The provision of properly constructed shade is essential in tropical latitudes. A Friesian cow confined without shade in the temperatures and humidities of the Persian Gulf will reach a stage of lethal hyperthermia within a few hours.

Ideally, shelters should have high insulating properties and should be extensive enough to eliminate much of the radiation from the sky. They should have a low radiation coefficient, and as a general rule should be at least 3 metres high. A large, leafy, non-deciduous tree approaches the ideal. Thatch from palm leaf, straw or grass is a very satisfactory material. Asbestos cement, galvanized iron or aluminium are less satisfactory but are usually the most commonly available materials. Their defects may be offset to a certain extent by making their surfaces reflecting (with silver or white paint), and by constructing the shelters properly. A highly effective but costly method of construction with these materials is to have a two-skin roof with an air gap of not less than 12 cm between the two skins. This air gap should have free access to the outside at the apex. If only a single skin of these materials can be afforded, the pitch of the roof should be made reasonably steep (not less than 20°); this will accelerate the upward movement of hot air layers adjacent to the undersurface. In addition, it should be as high as can be conveniently constructed, recognizing however that the extent to which a shadow moves is in direct proportion to the height of the shade. Thus when high shelters are designed, care must be exercised to ensure that the shadow is not cast outside the animals' quarters. Air gaps at the apexes of buildings for the exhaustion of hot air can be enhanced by constructing a flat or curved "lid" over the gap in order to produce a venturi effect.

In most areas of the Near East low wind velocities are a problem. But where high hot winds do occur, slatted walls or fences reduce the velocities considerably and interfere little with normal ventilation. The width between slats may be varied to suit prevailing wind speeds, but equal spacings are often a satisfactory compromise.

Feeding: Since a specific heat increment is associated with feeding, it is necessary to arrange feeding schedules to correspond with the cooler periods of the day — usually at night and in the early morning. Milking times should be organized in a similar fashion, since concentrates are normally offered at these times. The interval between milkings should be as wide as can be consistent with these factors.

Zero grazing systems in extreme climatic conditions have a decided advantage in that energy expenditure and heat production due to walking are kept to a minimum. However, if grazing is practised, the distance walked should be kept to a minimum and the grazing confined to the hours of darkness. Local conditions will of course dictate what diets are available, but experimental work has confirmed that more heat is produced from roughages for a given nutrient intake than from concentrates, and that succulent herbage is more readi'y ingested under high ambient temperatures than dry ligneous material.

Water: Water should, of course, be available at all times. The consumption of water by European cattle under heat stress may be more than twice that in their natural habitat.

Chilled drinking water is welcomed by stock. It undoubtedly reduces the heat load of the animals considerably, but the medium- and long-term effects of this practice on ruminant digestion have not been fully explored. This method of climatic alleviation must therefore be approached with caution.

Purebreds and Crossbreds

A detailed discussion of the crossbreeding programme in Thailand was given by Madsen and Vinther (1975). The programme was undertaken at the Thai-Danish dairy farm, which is situated 15° north of the equator and 230 m above sea level. The climate is tropical, with daily temperatures ranging from 19°C in the cool season to 38°C in the hot season, and with relative humidities of 65 and 95 percent in the dry and rainy seasons respectively.

When the project was initiated in 1962 no fixed breeding policy had been agreed upon. In the initial phase upgrading to Red Danish cattle was practised, but because of high mortality and disease incidence among purebred and grade cattle, back-crossing to the Sahiwal and Red Sindhi was adopted in the later stages. The foundation stock consisted of native cattle, improved native cattle (mainly white zebu from Burma), Indian milch breeds (Sahiwal and Red Sindhi) and Red Danish cattle.

Data were analysed using the least squares technique, thus correcting for environmental effects. Details of the method of analysis and of feeding and management practices have been provided by Madsen and Vinther (1975). That mortality was highest among purebred and high-grade Red Danish cattle, followed by Indian milch breeds. It was very low in the F^1 generation, presumably because of heterosis effects. Mortalities were roughly the same among the crossbred groups having 37.5, 62.5 and 75 percent genes from Red Danish cattle.

On the basis of the results, it was concluded that the optimum proportion of genes from Red Danish cattle for the environmental conditions prevailing in this experiment could be up to 75 percent but not more.

The age at first calving was high in the Indian milch breeds, whereas only minor differences were found between the other breeding groups. The calving intervals followed the same pattern as the figures for mortality, being very high for purebred Red Danish cattle, rather high for the Indian milch breeds, low in the F^1 generation, and with only minor differences among the other crossbred groups. The milk and butterfat yields in the first and second lactations increased almost linearly with increasing proportion of genes from Red Danish cattle. It was therefore concluded that the proportion of genes from Red Danish cattle should be as high as possible, but within the limits set by mortality and disease, i.e. about 60 to 80 percent.

First lactation yields were much higher for Red Danish cows imported from Denmark than for those born in Thailand. In the second lactations, however, the yields of imported animals were lower than those for animals born in Thailand. These results will be discussed later.

Purebreds in India

The Indo-Danish project is situated 30 km from Bangalore, 13° north of the equator and 880 m above sea level. The climate is tropical monsoon, with daily temperatures ranging from 10°C in the cool season to 38°C in the hot season, and with relative humidities of 63 and 86 percent in the dry and rainy seasons respectively.

A total of 18 bulls and 223 pregnant heifers were imported. These animals were used for purebreeding at the project farm. Male progeny were sold for breeding purposes to private farmers, but it has not been possible to trace the progeny of these bulls to any great extent. Nevertheless, it was possible to evaluate the performance of female crossbreds within the confines of the extension area of the Indo-Danish project. These results will be discussed in the following section.

Feeding and management at the Indo-Danish project were practised according to Danish standards as far as possible. The data were analysed using the least squares technique, thus correcting for environmental effects. The cows of the Red Danish breed were classified into three groups: cows imported from Denmark, cows born in India out of sires from Denmark (i.e. first-born calves from imported heifers), and cows born in India out of Indianbred Red Danish sires. The milk yields, butterfat yields and calving intervals for these three groups of cows. The difference of 326 kg milk and 11 kg butterfat between cows imported from Denmark and those born in India out of sires from Denmark is partly genetic and partly environmental. Genetic differences are expected because the imported animals were a selected sample of the population from Denmark, whereas very little selection was done among the calves born in India. Environmental differences may be caused by differences in feeding, management and/or climate during rearing.

The difference of 275 kg milk and 15 kg butterfat between cows born in India by sires from Denmark and those by Indian-born sires is a measure of the difference between the two sire populations. It is presumably caused by sires from Denmark being selected on the basis of progency tests, whereas sires from India were selected on the basis of pedigrees only.

Crossbreds in India

Milk recording began in villages in the Bangalore area in 1967 with the recording of 24 cows. In 1974 a total of 772 cows were milk recorded. Along with the milk recording scheme, an advisory scheme on feeding was operated. Thus cows were fed according to production. During the rainy season many farmers fed green fodder, while during the dry season most farmers fed only straw and concentrates.

Pedigree information was available for some of the cows in the milk recording scheme, making it possible to classify them into breeding groups: Red Sindhi, Red Danish×Red Sindhi. Red Danish × improved local cattle, and Red Danish × Hallikar. However, for most of the cows there was no pedigree information. Since most of them were crossbreds between European and local cattle, they were classified as improved local cattle. Data for all these groups were analysed using the least squares technique, thus correcting for environmental effects.

The 305-day yields of milk and butterfat and the length of calving intervals. The number of observations for the Red Sindhi is small, but the results coincide with those of previous investigations in that the milk and butterfat yields are low and the calving interval is long (Mahadevan, 1966).

It is assumed that the average proportion of *Bos taurus* genes in the group of improved local cattle was around 50. Few of these cattle were first crosses; thus heterosis effects are probably not important. The production records of this breeding group may therefore be considered as an estimator of the performance of cows which have approximately 50 percent *Bos taurus* genes but are unbiased by heterosis effects.

By contrast, the Red Danish × Hallikar and Red Danish × Red Sindhi breeding groups are expected to have their full heterosis effects. Assuming that the producing ability of purebred Red Danish cattle under village conditions is the same as at the Indo-Danish project, the estimated heterosis effects for the F_1 generation (Red Danish × Red Sindhi) are 16 percent for yield of milk, 30 percent for yield of butterfat and 16 percent for calving interval.

Thus crosses between Red Danish and Red Sindhi have performed better than might have been expected had there been no heterosis. This was also observed in crosses between these two breeds in Thailand. These results indicate that there may be a large specific combining ability between them on crossing.

A comparison of Red Danish × improved local cattle with improved local cattle shows that crossing with Red Danish cattle increased milk yield by some 457 kg and butterfat yield by 17 kg, whereas the calving interval remained virtually unchanged. On the assumption that improved local cattle have 50 percent *Bos taurus* genes, this difference would be a measure of the effect of increasing the proportion of these genes from 50 to 75 percent.

Imported Purebreds and Purebreds Born in the Tropics

In the Thailand data there was a decline in production from first to second lactation among imported heifers. The Indian data showed different results, but the increase in yield from first to subsequent lactations was much smaller than that normally observed in temperate countries. In India maximum production occurred in the second lactation, when correction for environmental factors was made.

However, the investigations coincided in regard to the decline in yield from the imported generation to the generations born in a tropical country. As noted earlier, the reasons for this are probably both genetic and environmental. It is not possible to specify how much of the difference is genetic and how much environmental, but it appears that first lactation performance was a biased estimator of the production performance of purebred Red Danish cattle in subsequent generations.

Optimum Proportion of Red Danish Genes in Crossbreds

It may be useful to consider environmental conditions in tropical countries as consisting of climate on the one hand and level of feeding and management on the other. In both Thailand and India feeding and management were at a high level and nearly up to normal Danish standards. However, the climate in Thailand was unfavourable, whereas that in India was not severe. Climate and feeding factors are summarize and compared with those of Denmark. The scoring in the table does not imply that the difference between the climate of Thailand and India is as large as that between India and Denmark; it is made for ease of reference only. The production levels attained by purebred cattle in India are lower than those normally obtained in Denmark, while those of Thailand were even lower than those of India. This pattern follows the scores for climate and feeding management.

In both Thailand and India milk and butterfat production was higher for purebred and high-grade Red Danish cattle than for intermediate crossbreds. Similar results were obtained in the United States by crossbreeding the Holstein, Jersey and Brown Swiss with the Brahman and Red Sindhi (Branton, 1966). But previous experience from tropical countries suggests that intermediate crossbreds had higher yields than high-grade animals (Rendel, 1972).

The most plausible explanation for this apparent discrepancy is that feeding/management was at a higher level in the Thailand, India and United States experiments than in those reviewed by Rendel (1972). If this interpretation is correct, feeding and management exert a greater influence on the productive performance of different genotypes than climate. This hypothesis is supported by the favourable performance of purebred cattle in Israel (Volcani, 1973). It is further supported by the fact that the season of calving had little influence on levels of production in Thailand and India. If climate had had a major influence on production, yields would have been low for cows calving in the hot humid season.

However, the reproductive performance of purebreds and high grades has not been satisfactory in either Thailand or India. The calving intervals were long, and in Thailand the percentage of abortions and mortality was also high. On the basis of the interpretations presented here, it would appear that delayed breeding is a reaction of the cow to relatively high production in an unfavourable climate. A similar phenomen is known to exist in temperate climates: cows are difficult to get in calf immediately after calving, when production is so high that the cow is in negative energy balance. Milk and butterfat yields have increased almost linearly with the increase in the percentage of genes from Red Danish cattle. This is contrary to several previous experiments with other European breeds, where intermediate crossbreds have had higher production than high grades and purebreds. This apparent discrepancy may be due to the high level of feeding and management in the present experiments.

However, mortality was high among high-grade and purebred Red Danish cattle under unfavourable climatic conditions in Thailand, and reproductive performance was not satisfactory in either Thailand or India.

Yields of imported animals were higher than those of animals born in the tropics. The reasons for this are probably both genetic (little selection in the tropics) and environmental (sub-optimal feeding/management and climate during rearing).

Cattle Breeding in Tropical South America

The tropical region of South America is characterized by an extreme range of climatic types. Of particular importance to its dairy industries are the modifying influence on temperature of the Andes mountains, which cover large areas of all the northern countries except Brazil, and the cooling effect of the Humboldt current on the climate of coastal Peru. The region's population of milked cattle includes cows of European breeds, distinct Criollo breeds such as the Colombian Costeño con Cuernos (horned Sinú), nondescript Criollo types, various zebu breeds and miscellaneous crosses between them all. The Criollo cattle have a history dating back to nearly 400 years on the continent. They are found at all altitudes, from the extreme lowlands to as high as 4 000 m in the Andes. In areas where climatic stress is not severe, most of the milk is produced by European breeds (now mainly North American Friesians) which are kept either on grassland in the mountains or in dry-lots around the cities. Many of them are imported or have been bred locally from imported semen, and they include animals of first-class genetic quality. As it is almost exclusively these cows that are performance recorded, statistics from the region relate mainly to them and are by no means representative of the total cattle population nor of the whole range of climatic zones. Apart from records from a few herds belonging to institutions or progressive farmers, there are relatively few published data for traits other than milk yield. Some examples from commercial farms, and it is evident that levels of performance which are fair by temperate zone standards are being achieved in these elite herds.

Research

It is to these herds that most research effort has hitherto been directed. In the field of breeding there is considerable interest in the comparative suitability of Friesian cattle of different origins for use in the region. From commercial herds near São Paulo, Brazil, it appears that the Dutch Friesian may have a slight advantage over the North American Friesian in areas where milk is sold according to solids content. This seems a rational basis for marketing throughout the region as malnutrition is widespread and transport difficult. Another comparison of Dutch and North American cattle is in progress under experimental conditions in Venezuela. In Peru, a comparison was made between the performance of locally bred Friesians and those imported from the United States and Canada (Sarmiento, 1970). The mean milk yields over the first six loctations were 4 924 kg, 5 125 kg and 5 263 kg for the three groups respectively, based on a total of 4 122 records. The mean age at first calving, 31 months, was the same for all groups and the interval between calvings about one month longer for the imported cattle. There is thus, predictably, little reason for the preference to import rather than breed herd replacements locally. With regard to imports, interesting results were reported recently from Maracay, Venezuela (McDowell, 1974). Of a group of heifers brought in from New York aged between 10 and 14 months, 60 percent were still in the herd five years later, while none of a second group, moved at the same time but already pregnant, survived. The two groups left progency of, respectively, 2.5 and 1.2 female survivors in the herd per head. This adds to the evidence from elsewhere in the tropics that better adaptation results if heifers are imported well before breeding age.

There is also interest in the region as to whether dairy bulls should be selected specifically for the various ecological zones. So far the only published evidence comes from farms in Ecuador, ranging in altitude from 2 250 to 3 400 m, with mean annual (temperatures between 10 and 17°C (Roman, 1970). The sire × herd interaction accounted for only 2.9 percent of the total variation in milk yield. These results might have been more important if farms in the extreme lowlands had been included in the study, but they match with existing evidence that it is not necessary to select sires for the conditions under which their daughters will perform. Thus there is still no evidence that the best proven sires from temperate zones are not also the most profitable for tropical use.

Special Problems

Two areas in the region present special problems to the milk producer: the extreme lowlands, which include vast tracts of jungle sparsely inhabited by cattle or man, and the Andean highlands.

Highland Areas

In the highlands, European breeds and Criollo cattle are kept up to altitudes of about 4 000 m. Of the European breeds the Friesian and the Brown Swiss are the most popular, and the latter is traditionally supposed to be less susceptible to altitude sickness (brisket disease) than the former. However, problems with the Friesian seem to be less serious if they are really well fed and managed, and at least one Friesian herd is kept successfully as high as 4 000 m in Puno, Peru. The difficulties of feed supply are probably the most serious obstacle to increasing milk production. Concentrates are generally too expensive to use freely, and seem likely to remain so because of high transport costs. Present trends are to develop varieties of the improved temperate grasses, legumes, and forage crops such as oats. Under such a grazing regime at Huancayo, Peru (altitude 3 200 m), a mixed herd of Friesians and Brown Swiss cattle produced an average yield of 10 kg milk daily when stocked at five per hectare.

Lowlands

Milk production in the lowlands is generally backward. Perhaps the only example of a thriving local industry is that of the Maracaibo region of Venezuela, where Criollo Lechero and crossbred (mainly with Brown Swiss) cows produce average yields of 1 500–2 000 kg per lactation, principally from grazing. The performances of Criollo, Costeño con Cuernos, Friesian, Brown Swiss and crossbred cattle are being compared at Turipaná in northern Colombia, where the annual mean temperature is 28°C and the average relative humidity 83 percent. From results so far published, it appears that first-cross cattle gave as much milk as the exotic breeds but were easier to get in calf. The Criollo cattle produced less when milked alone than with their calves at foot and, although their conception rates were good, the relatively low proportion of animals pregnant again by 100 days after calving was due to difficulty in detecting their oestrus periods (Huertas, 1972).

Work in Peru suggests that the costs of preparing and maintaining jungle land are so high that cattle enterprises, at least by themselves, are of doubtful profitability. The only recent published dairy records from the region refer to Dutch Friesians taken at an average age of 19 months to Pucallpa, where the annual mean temperature is 25.4°C, with mean maximum and minimum values of 30.8°C and 18.4°C. The cattle were kept mainly on pastures of *Hyparrhenia rufa* and, because of prohibitive costs, only a little concentrate feed was given at milking time. Their performance (Meini, 1973) is summariz. Their-reproductive performance was typically poor and is overestimated by the data shown, as conventional statistics such as these do not take into account the number of cows which do not conceive or calve. Here, long calving intervals were due to anoestrus under a system of inspecting the cattle twice daily for heat, irregular oestrus periods and repeat services (AI). There is limited preliminary evidence from the same station that Criollo cattle (of highland origin) and Brown Swiss may be more productive than the Friesians because their yields of milk are similar but their reproductive performance is better. For instance, the mean calving interval of Criollo cattle at the station is 13.8 months

Improvement Through Breeding

Improvement through breeding has been slower than it need have been throughout the region, and the reasons for this seem worth examining. First, there is widespread ignorance at government and farm levels as to what genetic quality for milk production consists of and how it is measured. The fact that AI and recording services exist in many parts of the region is frequently considered to be sufficient to solve all problems; also, the bulls used are often of inferior quality. An illustration of this comes from a study of Friesian sires used through AI in the Lima area in 1968. It was found that 42 percent of the bulls had negative progeny test ratings, that is, were sires whose future daughters would be expected to produce less milk than their contemporaries (Vaccaro *et al.*, 1968).

Costs are inflated and genetic progress hampered by the continued importation of bulls which, were they of first-class quality, would probably not be for sale. Instead, bulls for local use should be bred from the semon of the best foreign proven sires and selected cows. A great deal of emphasis is commonly given to "type," which undoubtedly does have a high commercial value throughout the region, whereas greater publicity should be given to the fact that selection for physical characteristics does little or nothing to improve dairy productivity.

Second, selection is hindered in many parts of the region by a shortage of females. Instead of resorting to frequent importation, longer term benefits would result from investing in heifer calf subsidies and relevant technical assistance. Poor reproductive efficiency is an important contributing cause of the scarcity and, in this connexion, interesting results emerged from a survey carried out by Zemjanis and Sanint (1963) on cows on 32 Colombian farms. Most of the 4 122 cows in the sample were kept in the highlands around Bogotá, where heat stress is not a problem.

An average of 3.3 services were required for conception, but ranged from 2.0 where commercial (mainly frozen, imported) seme was used, to as many as 11.5 where herd bulls were used through AI. Furthermore, the difference between results obtained when the same bulls were used naturally and artificially emphasized the need to improve semen handling and insemination techniques. It is also significant that as many as 47 percent of the sample were classified as anoestrus and that as much of this was thought to be due to faulty detection of heat. In the lowland tropics particularly it seems likely that heats would be missed unless cattle were inspected frequently: a study at Maracay suggested that a minimum of two (at 0600 hours and 1700 hours) and up to four observations daily may be necessary (Fenton *et al.*, 1972). The cattle in question were imported Friesians, and it was found that genuine anoestrus tended to persist as long as the animals were losing body weight after calving. Best conception rates resulted from breeding the cows within 12 hours after the detection of oestrus. By measures such as improved nutrition and culling, it was possible to reduce the average number of "days open" in this herd from 552 to 220 over a period of 18 months (McDowell, 1974).

Other Considerations

It may be argued, however, that genetic progress has a comparatively minor role to play in the overall improvement of the region's dairy industries. The other ingredients are well known: they include better feeding and management, disease control, price and quality regulations, credit services and marketing facilities. What has happened in the past is that improvement efforts have too often been insufficiently comprehensive or, because of political changes, too short-lived. There is nevertheless a genuine lack of information about genotypes of cattle for profitable lowland farming systems.

The lack of useful performance data from the lowlands is notorious, and quite incommensurate with the local resources and foreign aid already spent there on dairy work. One reason is that policies have usually changed too quickly to permit the accumulation of sufficient data for the purposes of genetic study. Also, policy decisions are generally made away from the region and on the basis of insufficient knowledge of the local environment. This is especially true where the lowland experiment stations are difficult of access. Such stations have the additional problem of attracting and keeping the highly trained personnel that tropical animal science requires, as facilities for family living are often very limited. Dairying research in the lowlands has been insignated because large areas of land are unexploited and urban people need dairy produce, not because dairying is a significant part of local agriculture. In fact, few lowland areas in the region have any tradition of keeping cattle. Experiment stations have already seen how this fact affects the success of research programmes, and it raises serious questions about the type of dairying systems, and consequently breeding programmes, which are really applicable.

Despite the difficulties which face them, governments in the region are actively interested in improving milk production. Peru, for example, which now imports slightly over 50 percent of its requirements, expects to be self-sufficient in milk products by 1980. The frequent political changes which characterize the region make progress difficult, particularly in animal breeding, but real hope lies in the unprecedented number of highly trained professionals educated at home or returning from abroad, provided that they are increasingly involved in the planning and implementation of national livestock policies.

The Tropical Adaptation of Beef Cattle

Introduction of Tropical Cattle

Whereas most tropical countries have indigenous *Bos indicus* cattle and look to the introduction of *Bos Taurus* to improve productivity, in northern Australia it has been necessary to introduce tropically evolved cattle. The approaches to the target of adapted, productive cattle from opposite starting points are instructive. In each case the focus of interest tends to be on the novel qualities to be introduced, sometimes to the neglect of those already established which are taken for granted.

Quarantine regulations have limited the sources and extent of importations of tropical cattle. The early entry of a few zebu cattle and of buffalo and banteng had little impact, the former being dissipated and the latter isolated as feral populations. In 1933, 19 Brahmans were imported from the United States and distributed among a number of herds. These were followed in the 1950s by further imports of Brahmans and Santa Gertrudis, by a small number of Africanders (also from representatives of the breed in the United States) and by Sindhis and Sahiwals from Pakistan. The latter were intended primarily for dairy breeding, but the use of Sahiwals in beef breeding has been explored.

Belmont Programme

The breeding programme at Belmont, the field station of CSIRO's Tropical Cattle Research Centre at Rockhampton, started in 1954 with representatives of Brahman, Africander, Hereford and Shorthorn breeds. This centre was set up for the study of inherited factors controlling adaptation and economic performance under northern Australian conditions. After initial matings with all combinations of breeds, the following are the main lines established and carried forward, currently, to the F_4-F_5 generation:

Africander cross: half Africander, quarter Hereford, quarter Shorthorn; Brahman cross: half Brahman, quarter Hereford, quarter Shorthorn; Hereford-Shorthorn: half Hereford, half Shorthorn.

These are referred to respectively as AX. BX and HS. In these and other subsidiary lines, a breeding herd of about 900 cows is maintained. They and their progency are grazed on natural and improved pastures representative of the area and exposed to the normal stresses of the environment. Performance is compared principally under these conditions, but some growing stock are studied more intensively in confinement.

Performance Attributes

Fertility

The calving percentages represent breed means accumulated over a number of years, corrected for the effects of age, lactational status and year (Seebeck, 1973a). These results were all obtained from a short annual mating period of seven weeks, with young (2-year-old) bulls and with singlesire matings (one bull to 30–35 cows). All three factors lower the level of fertility realized and emphasize differences in reproductive efficiency. In matings of $F_1 \times F_1$, the differences between breeding lines were not significant, although both AX and BX were more fertile than HS. In subsequent generations, AX maintained high fertility, HS dropped slightly, and BX fell dramatically; the breed differences are highly significant. Reciprocal matings between the AX and BX lines, made simultaneously with straight matings of each line, have shown that the differences is expressed in both males and females.

Within the BX and HS lines, cows that are progeny of different sires differ significantly in fertility, with a heritability of 22–25 percent (Seebeck, 1973a). There is scope for improving fertility by selection, which would be enhanced by identifying the underlying genetic factors. Toward this end, the significance of morphological abnormalities of the female reproductive tract, semen abnormalities, serving performance, fertilization rates and postpartum anoestrus is being defined, and the endocrine bases of some of these differences are being elucidated.

Growth

The mean body weights of calves of the three breed lines. These represent 500–600 female calves of the F_2 and F_3 generations born over a period of five years. The breeds did not vary significantly in birth weight, but thereafter the differences were highly significant. At 18 months, BX were 21 percent heavier and AX 16 percent heavier than HS. These differentials are somewhat lower than in F_1 progeny, the decline from F_1 to F_2 being greater in BX than in AX, and more in birth weight than in subsequent weights.

In breeding cows, mature weights are very similar in the different breeds, but they are approached at different rates and differ in seasonal stability. During a drought, weight changes of nonlactating pregnant cows from February to October were: AX-8.7 kg, BX +5.8 kg, HS-33.0 kg. However, the changes of lactating pregnant cows were all similar at about-33 kg, the BX thus being most affected by lactation (Frisch, 1973a).

Mortality

The mortalities of breed groups in calves of F_2 + generations from birth to 15 months and in adult cows (Frisch, 1973b). At all stages, mortality was least in AX and greatest in HS. Comparison were similar in F_1 animals except that perinantal deaths were high in zebu-cross calves, especially those born to young British-breed cows. The mortalities of adult cows, averaged over a number of years, rose in two drought years to 5.6 percent in HS, 2.0 percent in AX and 1.5 percent in BX.

Carcass Composition

The yield of carcass (dressing percentage) is higher, by about 2 percentage units, in BX than in AX or HS steers (Hewetson, 1962). The zebu-cross carcasses are leaner under some conditions but breed differences in fat content depend on stage of growth and plane of nutrition. Fat distribution shows some breed differences (Seebeck, 1973b). Distribution of muscle weight also varies, with AX better developed in muscles surrounding the spinal column and BX better developed in muscles of the upper hindquarter. In general, differences in yeild and quality of meat at a given liveweight, although significant in indicating the potential for genetic improvement, are of minor immediate importance in most market situations.

Pure Zebu Performance

The foregoing compares the performance of *B. indicus* × *B. taurus* halfbreds with a line representing their *B. taurus* parents. Directly comparable data on the performance of the zebu parental breeds are less extensive, and it must be remembered that representatives of these breeds have only a small genetic foundation in Australia. Nevertheless it can be stated that compared with the crossbreds the Brahmans and Africanders have lower growth rate, lower fertility and higher juvenile mortality, but lower postweaning and adult mortality.

Genetic Adaptations to Components of the Environment

The preceding comparisons of performance were recorded in a specific field environment. Obviously they would not be identical in all environments. Adequate definition of the physical and biological environment as its affects cattle is difficult to express in absolute terms, and it is more meaningful to identify and quantify elements of the environment in terms of their comparative effects on animal performance. This gives a perspective of elements limiting performance, the importance of genetic differences in response to them, and the genetic strengths and weakness of particular breeds.

The effect of heat in a field situation has been estimated by clipping the animals; coats (Turner, 1962). Keeping Herefords clipped increased their growth during the six warmer months of the year by 13 percent. As clipping only partially relieved heat stress and lowered rectal temperatures to less than those maintained by zebu crosses under the same conditions, this is a minimum estimate of the importance of heat to a susceptible breed. Elementary parameters of the thermal environment in which these results were obtaine.

The superior heat tolerance of zebu crosses represents a significant advantage under warm conditions. Its main determinants are a sleek coat, high sweating capacity and low body heat production. Coat type is a good index of tropical adaptation in temperate breeds; it is related to sweating capacity (Schleger and Bean, 1971) and has a high heritability and genetic correlation with growth rate (Turner and Schleger, 1960).

Parasites and Diseases

Breed differences in susceptibility to parasites have been studied by comparing responses to parasite control. In herds containing representatives of three breed groups, all run together at pasture and exposed to normal field infestations of the cattle tick (*B. microplus*) and gastrointestinal mematodes, one third of the animals were dipped twice or three times weekly to control ticks, one third were treated with anthelmintic every two or three weeks to control worms, and one third were left untreated. Gains of each treatment group of each breed in two different experiments show that HS were profoundly affected by both ticks and worms whereas BX were unaffected by either and AX were somewhat affected by both. Genetic tick resistance, gained by breed selection, crossbreeding, and within-breed selection, is an important contributor to performance in endemic areas and the most efficient means of countering the tick. Genetic differences in susceptibility to gastrointestinal nematodes are obviously of similar importance. A criterion of helminth tolerance, as applicable for selection as are counts of mature ticks in selection for tick resistance, remains to be developed.

Infectious Keratocunjunctivitis (pinkeye) is an affection common everywhere, and is usually considered to be of minor nuisance value. It has been shown (Frisch, unpublished) that affected animals have markedly reduced growth rates and that there are major breed differences in susceptibility.

These are a few examples of the existence of genetic differences in susceptibility to diseases and parasites. Genetic solutions must be considered as economical and ecological alternatives to other control measures.

Feed Utilization

Genetic differences in the growth response of animals subjected to stresses such as heat or parasites are expressed through some aspect of feed utilization. Susceptible animals suffer depression of feed intake or effects on digestion or metabolism (e.g., Seebeck *et al.,* 1971). There are other inherent differences in feed utilization independent of environmental stresses. Voluntary feed intake per unit of liveweight is comparatively low in Brahmans, but they also have a lower maintenance requirement at a given liveweight (Frisch and Vercoe, 1969). It remains to be seen whether a low maintenance requirement, which is a component of efficiency and a great advantage under conditions of feed shortage, can be combined with a high feed capacity which promotes gross efficiency under conditions of abundant feed. For the present, a low maintenance requirement is of overriding importance where there is risk of seasonal feed shortage.

There is a tendency for zebu crosses to have slightly higher digestive efficiency than British breeds (Moran and Vercoe, 1972). Another genetic difference of potential importance for maintenance on low nitrogen diets is in the recycling of urea to the rumen. Zebu crosses maintain higher blood urea under certain conditions, and animals with low water intake and urine volume, advantageous in an arid environment, also conserve urea (Vercoe, 1967).

Genetic differences in the commitment of nutrients to fat or protein deposition, and in the relative depletion of fat or protein stores during weight loss, affect the efficiency of meat production.

Environment or Genotype

In relation to the target of improving animal productivity there tend to be two schools of thought. One favours modifying environments to accommodate animals of the highest productive potential, the other favours neglecting any scope for ameliorating the environment and accepting the productivity of genotypes which withstand its rigours. Either is a prejudgement and neither alone gives optimum solutions. Freeing the environment of all limitations, in climate, disease, parasites and nutrition, is in some cases not feasible and in many not economical. Any managerial measure relying on inputs (buildings, mechanization, fertilizers, pest control) must be evaluated in terms of cost and benefit. Cost must be viewed not only in immediate monetary terms but with a long view of true costs of inputs such as fossil fuels and nonrenewable resources, and side effects on ecological balance. Cost-benefit evaluation of managerial inputs may be seriously misleading if limited to one genotype, ignoring genetic differences in response. Conversely, genetic solutions may be inefficient where scope for effective adaptation is limited and an environmental element can be transformed by minor inputs. To optimize productivity it is necessary to study both factors: modification of components of the environment and genetic differences in response to them. The examples of such investigations at Rockhampton given in this articles have been more comprehensively reviewed by Vercoe (1974).

Using Large Muscular Breeds to Improve World Beef Production

The beef consumed throughout the world comes from three main types of enterprises:

1. Dairy herds where females are milked and meat is a by-product of milk production.
2. Single suckler herds (cow-calf operations) where specialized beef breeds (which tend to be small and fat) are used in environments that are favourable to beef cattle raising (i.e. grassland areas with a temperate climate).
3. Single suckler herds where, because of unfavourable conditions of climate, disease, and/or nutrition, local cattle adapted to adverse environments are utilized.

The breeds used in these three situations generally lack the muscular growth potential needed to meet the modern consumer's demand for lean, tender meat. This accounts for the efforts made in the last 20 years to improve beef cattle for meat production, principally by encouraging the use in crossbreeding of breeds with a high muscular growth potential. These are primarily continental European breeds raised in favourable environments and selected according to objectives (work and meat) that give full scope for the manifestation of this trait.

After an initial over-optimistic phase in which the muscular growth potential of "exotic" breeds generally showed spectacular development in the first generation crosses, there appeared various shortcomings in the fitness components of the crosses (particularly females). The problem of optimum utilization of exotic genetic material with a highly developed musculature was then approached with a view to increasing meat production when starting from an indigenous female foundation stock. What, in this situation, is the best crossing scheme to use? The answer to this question depends on several considerations concerning the genetic material:

1. The relative importance of hybrid vigour and of additive gene effects on the desired trait. This will determine the method of crossing to use, i.e. systematic crosses for the market or creation

of new breed types. The first method, in contrast to the second, leads to maximum manifestation of hybrid vigour.

2. The importance of direct effects compared with maternal effects of the genes determining a given trait. In this connexion, the independence (and even antagonism) which seems to appear statistically between breed types for these two kinds of effects can lead to the use of "muscular breeds" to produce calves entirely for slaughter (terminal cross), or from which the females will be used for breeding for only a limited number of reproduction cycles or not at all. This leads us to determine:

   a. A paternal value or component of breeds, for one or more traits; this appears when comparing the meat value of the progeny of bulls of these breeds crossed with a given type of female.

   b. A maternal value or component of the same breeds, i.e. that manifested for one or several traits influencing production efficiency when we compare females of these breeds mated with a single breed of bull.

3. The adaptation of the breeds concerned to the production environment and, more specifically, to climatic, health, and nutritional factors comprising that environment. What proportion of genes of local breeds should be kept for adaptation to one or another natural or manmade stress (for example, suckling or artificial feeding) in introducing improved breeds?

The research conducted in France can provide some information on these aspects because:

- Many of the "muscular breeds" are native to France and are widely exploited there; one can therefore compare more uniform and more representative samples of animals from the populations existing in France than from those used in experiments carried out in the new user countries.

- The French breeding environments are fairly representative of the various beef breeding conditions mentioned above, i.e. dairy cattle breeding; beef cattle breeding in favourable environments (areas of central France populated by Charolais and Limousin), or unfavourable environments, i.e. mountainous or subhumid areas of central and southern France where the animals must be specially adapted to prolonged nutritional deficiencies in the summer or winter.

- The remarkable development of artificial insemination in France (70 percent of females inseminated, 50 percent of them belonging to suckler cow herds) affords a useful structure for setting up experimental selection or crossing experiments and, more generally, for cooperation between researchers and livestock breeders' associations in the field of genetic improvement.

Improvement of Meat Production from the Dairy Herd

The process of selection for milk production leading to the gradual replacement of less productive breeds, strains or lines by others highly specialized for milk production makes it possible to rapidly increase meat production independently of milk by crossing the lower yielding cows (30 percent on average) with bulls of the beef breeds.

In general we are dealing with small-or medium-sized farms using artificial insemination. First generation crosses principally comprise heifers that are managed intensively, inseminated from a bull of a small beef breed, and probably slaughtered after an early first calving. This establishes a fattening unit of "once-bred" heifers that is partly self-replacing (FAO, 1972). However, when the milk yield is low and the farm large, milking may be given up and cows bred to beef bulls and used for suckler calf production (e.g. in the United Kingdom and Ireland).

In any case, as the aim generally remains a terminal type cross, the tendency will be to choose beef breed males on the basis of their paternal components or values mentioned above, as reflected by the calves' potential for muscular growth.

In line with this approach, the French insemination centres concerned, in liaison with the National Institute for Agricultural Research (INRA), have been developing programmes since 1955 that are designed to create. select and compare beef strains for use in terminal crossing (Centre national de recherches zootechniques, 1972). The strains studied have consisted of:

1. Registered animals (Charolais, Limousin, Blonde d'Aquitaine) of normal type, as distinct from the type with muscular hypert ophy or double-muscling. The percentage of isolation of strains in the registered population varies according to breed, being higher in Blonde d'Aquitaine and Limousin, where the selection schemes for terminal crossing are tending to lead to completely isolated strains.

2. Breed crosses between registered animals of normal type. The breed combinations are designed to bring together more compact genetic types having higher growth potential (Charolais, Maine-Anjou) with longer bodied and taller breeds (Blonde d'Aquitaine, Limousin) which for a given weight at birth have a lower frequency of calving difficulties.

3. Crosses between animals with muscular hypertrophy. Unlike the two preceding categories, which are selected by the insemination cooperatives, this operation is carried out by INRA on its Carmaux farm which has two herds of female crosses: Charolais X Blonde d'Aquitaine and Maine-Anjou X Limousin or Blonde d'Aquitaine. In this and the preceding case, the schemes tend toward the production of isolated populations. Males of all these strains are first subject to individual within-strain selection and are then simultaneously progeny tested in comparison with a national control lot (Charolais, Limousin or Blonde d'Aquitaine) according to the value of their crossbred progency out of dairy females; the female base is progressively becoming Friesian. The crossbred calves, initially raised for veal and recorded on the farm, are increasingly being recorded at a testing station, half being slaughtered at 15 months and half at 18 months, thus giving a comparison of strains based on different ages at slaughter (Centre national de recherches zootechniques, 1972).

Although the initial objectives of bull selection by both performance and progeny tests were designed to improve only the muscular growth potential of the animals or their progeny, other criteria have gradually been introduced into the index:

a. Efficiency of feed utilization (independent of growth rate).

b. Meat quality (some components of which can be detrimentally affected by increase in muscularity).

c. Ease of calving (direct effect). Indices restricting genetic progress in birth weight have been constructed, based on the males used for crossing; moreover, semen distribution can be based on the bull's birth weight index on the one hand, and on the dairy females' ease of calving on the other. If little or nothing is known about the ease of calving, the female population can be classified by age and breed and according to size of pelvic opening or weight, i.e. estimated size for each age and breed class based on a small sample of individuals.

In fact, it is becoming increasingly difficult for a single strain to meet all the objectives sought in a terminal cross. These objectives vary widely according to:

a. Calving age and ease of calving of females. Some insemination cooperatives have therefore been involved in the formation of strains known as "maximal" (no limitation on increase in weight at birth) and "optimal" (strict limitation of the increase), the latter being mainly confined to early-calving heifers. Such an approach should be able to provide for the most rational utilization of the pelvic passage which, in the long run and with only a single calf, in the female must constitute a basic limiting factor to beef production.

b. Type of production of crossbreds (especially females), i.e. milk X meat. In this connexion, increased muscular development within strains or between strains is generally accompanied in crosses by a delay in body maturity. Thus the use of double-muscled bulls on Friesian cows delays by one to two months the age at which the carcass of the crossbred animal shows a given fat percentage. This can be a considerable advantage in the case of crossbred females which under an intensive management system (slaughter at about 14 to 16 months) can thus reach slaughter weight without getting too fat.

On the other hand, females produced by doublemuscled bulls will show a reduction in fitness (sexual maturity, ease of calving), so that such bulls are less used in systems involving slaughter of females after an early calving (Bibe *et al.*, 1974).

The field is thus open for competition between beef cattle strains designed for crossing with dairy females by insemination. Comparative data should now be refined according to the type of female being crossed and the type of meat production. To be effective, an analysis of the results will make it increasingly necessary to coordinate experiments because it suggests an easy way of increasing meat production independently of milk in industrialized countries which have mainly dairy cattle bred by artificial insemination.

Improvement of Meat Production from Specialized Beef Herds

In systems where female progeny are normally kept for breeding, numerous shortcomings of "muscular breeds" in extensive production systems have appeared, particularly in the fitness components. These observations are logical enough, considering the artificial environment in which the animals have sometimes been maintained in small-and medium-sized herds, i.e. calving supervision and assistance, control of heat periods, supplementary feeding of calves (nurse cows), and housing in winter and during the night in summer.

The results of experiments in France and other countries reveal a genetic antagonism between the calves' potential for muscular growth and female fitness components; in the case of traits depending on both the calf's and the dam's genotype (e.g. weight at weaning), this antagonism seems to involve the offspring's growth genes on the one hand and the genes controlling the dam's milk production and maternal behaviour on the other. This conclusion emerges clearly from the comparisons that can be made (a) between French breeds of variable size and muscularity raised in specialized beef herds; (b) between Charolais strains selected in different countries (the differences in musculature between these strains, e.g. between Brazilian Charolais and United States Charolais specifically selected for terminal crosses, are greater than those found on average between animals of different breeds) and (c) between progeny of bulls. Here, the French results confirm those obtained in the United States on the antagonism between the direct and material effects determining weight at weaning.

Such observations lead us to seek the best equilibrium in genetic structure, either by selection within breeds or in crossbred populations, or through systematic crossbreeding. The first method will seek to exploit mainly the additive and epistatic effects of genes. The second, which is more difficult to implement, will also involve dominance effects, and above all, complementarity between the calf's and the dam's genes which will depend, in particular, on the order of the introduction of breeds in the crossing schemes. The search for such equilibria implies an analysis of the genetic variability between and within "muscular breeds" and smaller beef breeds (Hereford, Aberdeen-Angus). The experimental and improvement programmes undertaken in central France in the areas where specialized beef production predominates have been carried out with precisely this aim.

Studies on between-breed variation: A factorial plan of crosses between samples of animals representing the registered populations of the three main French beef breeds (Charolais, Limousin, Maine-Anjou) was set up on the Bourges experimental farm, which also has a purebred Hereford control herd. For practical reasons, the females are zero-grazed. The results of the first four years of operation present can be summarized as follows (Menissier *et al.*, 1974, 1975):

- Under the conditions of this test the Limousin breed presents the most favourable maternal components for measurable productivity traits. It also gives the best results during the fattening phase (paternal and maternal components) when the animals are slaughtered at 18 months. On the other hand, for earlier slaughter (15 months), all three breeds give comparable results.

- Reciprocal crosses between Limousin and Maine-Anjou show excellent complementarity in calving fitness as well as weight at weaning and carcass value (body composition).

- Heifers and young cows of the Charolais and Maine-Anjou breeds in particular show a high frequency of calving difficulties. In this dystocia, the relative importance of paternal and maternal components (size of foetus and pelvic opening) varies according to breed and cross. We might consider decreasing these difficulties through reducing the adult size of breeding females by applying crossing schemes in which the females are divided by age into two groups. Thus, for their second and third calvings cows are mated in a rotational crossing system consisting of alternating small breeds which complement each other for the different components of fitness, including ease of calving; all the female offspring are kept for breeding. Older cows are mated to larger bulls with greater muscular development, and all male and female offspring are slaughtered. Such a system of course eliminates the possibility of female selection, but the generation interval is shortened considerably.

Studies on within-breed variation: Since selection for fitness proved necessary in the case of the "muscular" French breeds, it was essential to find more efficient methods than natural selection, which was considered adequate under extensive conditions (e.g. in North America and Australia) with breeds of British origin. This is the object of selection schemes for bulls of beef breeds used for artificial insemination in France in specialized beef breeding areas (Foulley and Menissier, 1974). The schemes include in particular a station test of female progeny — a comparison of 20 female offspring per bull, from weaning to 30 months, which are fed intensively so that they can be calved down at 2 years. Although this system is a considerable departure from more extensive methods adopted in many countries and covers only a short phase of the animal's reproductive life, it was considered *a priori* to be effective for "muscular breeds."

Sardinian, Modica, Brown Swiss and first-cross cows with Charolais and Piedmont in the experimental beef herd at the Sardinian Zootechnical and Dairy Institute at Foresta di Burgos, Sardinia.

As seen earlier, the basic handicap of their maternal components influences the traits appearing early in the females' life (sexual precocity, ease of first calving). A positive correlation may be presumed between the traits related to the first reproduction cycle and those concerned with later cycles. In this connexion, the greater variation appearing at a young age may provide a better assessment of the genetic variation in the adult female in the same traits (e.g. fertility), even with a small population.

Improvement of Meat Production from Beef Herds in Adverse Environments

The limited world feed supply and the high cost of beef production in fertile areas suited to intensification are leading to increased use of poor grazing areas (arid and mountain zones) and of the by-products of intensive cultivation by beef cattle which, with sheep and goats, are the only animals able to exploit profitably these areas and products. But it is necessary to find the best way of improving these potentials of local female populations adapted to the environments by crossing with males of beef breeds. We must also determine the possible use in such environments of crossbred females (milk X meat) from dairy herds in nearby intensive environments such as irrigated lowland areas associated with arid mountain areas, e.g. in the Mediterranean basin.

The research undertaken in cooperation with the Sardinian Zootechnical and Dairy Institute on hardy breeds of southern France and Sardinia provides useful information on this question in areas characterized by more or less prolonged undernutrition.

When suckling cows are maintained in protected environments in which they are given feed supplements during periods of undernutrition, a two-stage crossing consisting of two successive crosses of local females with meat breed males (Charolais or Blonde) seems to be the optimum solution. Experiments on this were undertaken in southern France with the Aubrac and Gasconne breeds. In the long run this formula is more effective than upgrading the local breed to the meat breed. But when local environmental conditions cannot be changed (as in the Sardinian experiments), only a first generation crossing of local females with beef bulls will make local beef production more profitable. This is partly due to the relatively good maternal fitness of the local breeds that were exploited in the past for milk, meat and draught.

Double-muscled bulls and their female offspring have a reduced fitness and poorer adaptation which often make their use in breeding less effective in adverse environments. This also generally applies to crossbred females, i.e. offspring of dairy cows by beef bulls.

All of the results stress the importance of limiting the introduction of new genetic material in adverse beef cattle breeding environments. Preliminary studies should be undertaken to set the limits to crossing of local females with males of improved breeds. They should cover a range of environments representing present conditions as well as economically and humanly foreseeable environmental changes made by man to improve health and nutrition. As we have shown, they are likely to lead to solutions requiring the maintenance of a large sample of the local breed.

In more general terms, it is necessary to establish the interaction existing between genetic types of beef cattle and their environments. However, direct analysis of these interactions as traditionally practised (i.e. simultaneous comparison of several genetic types in several environments) requires a much too costly experimental set-up often prepared on too vague a basis. In fact a given environment represents the synthesis of disease, climatic and nutritional stress. It would be advisable to analyse thoroughly the genetic differences existing between beef cattle populations with reference to each basic type of adaptation or stress. There would then be a better understanding of the interactions discovered, and the field of application of the experimental results could be expanded. The initial work done in France and Sardinia with this approach has led to some useful conclusions:

- The higher productivity of local females under the extensive management conditions of Sardinia is related to their apparently greater ability to maintain their body weight and that of their foetus and their milk production during or after a prolonged drought period (Casú *et al.*, 1975).
- The ability of suckler cows of beef and hardy breeds to maintain their body temperature when subjected to a controlled heat stress seems to be related to their musculature.

Such tests could, if necessary, be used to improve the sampling done at present in temperate countries to select breeds and genetic types for use in crossbreeding to improve maternal performance of beef cattle in adverse environments.

Conclusions

Genetic improvement of the potential for muscular growth, which in most countries and situations seems to be regarded as the quickest way of increasing meat production and adapting such production to consumer tastes, presents considerable limitations which must be understood. In fact it is more or less closely associated, depending on the situation, with a lowering of fitness and adaptation traits of both males and females. This observation gives rise to several considerations:

- We will generally be led to seek a balance between growth and adaptation traits, such a balance being determined by the conditions of the natural or artificial environment with which we are dealing.

- Crossing between breeds showing complementary traits (small female X large "muscular" male), which also enables hybrid vigour to be exploited, will become a necessary part of any plan for improving meat production.

- It is therefore becoming increasingly important for the producer to maintain the variability of existing genetic material. This will be achieved mainly by exploring in a more thorough and dynamic way the possibilities of exploiting this material in environments possessing the feeds and breeding types to which it is particularly suited. In this respect, we can only regret the scarcity of our information and the excesses that have led to the wide application of results obtained in special and frequently favourable intensive environments to situations where beef production can often be operated profitably only in marginal areas or on by-products, or in some cases must itself be only a by-product of dairy production.

Spoilage and Fermented Milk Products

When raw milk is left standing for a while, it turns "sour". This is the result of fermentation, where lactic acid bacteria ferment the lactose inside the milk into lactic acid. Prolonged fermentation may render the milk unpleasant to consume. This fermentation process is exploited by the introduction of bacterial cultures (e.g. Lactobacilli sp., Streptococcus sp., Leuconostoc sp., etc.) to produce a variety of fermented milk products. The reduced pH from lactic acid accumulation denatures proteins and caused the milk to undergo a variety of different transformations in appearance and texture, ranging from an aggregate to smooth consistency. Some of these products include sour cream, yoghurt, cheese, buttermilk, viili, kefir and kumis.

Pasteurization of cow's milk initially destroys any potential pathogens and increases the shelf-life, but eventually results in spoilage that makes it unsuitable for consumption. This causes it to assume an unpleasant odour, and the milk is deemed non-consumable due to unpleasant taste and an increased risk of food poisoning. In raw milk, the presence of lactic acid-producing bacteria, under suitable conditions, ferments the lactose present to lactic acid. The increasing acidity in turn prevents the growth of other organisms, or slows their growth significantly. During pasteurization however, these lactic acid bacteria are mostly destroyed.

In order to prevent spoilage, milk can be kept refrigerated and stored between 1 and 4 degrees Celsius in bulk tanks. Most milk is pasteurized by heating briefly and then refrigerated to allow transport from factory farms to local markets. The spoilage of milk can be forestalled by using ultra-high temperature (UHT) treatment; milk so treated can be stored unrefrigerated for several months until opened. Condensed milk, made by removing most of the water, can be stored in cans for many years, unrefrigerated, as can evaporated milk. The most durable form of milk is milk powder, which is produced from milk by removing almost all water. The moisture content is usually less than five percent in both drum and spray dried milk powder.

Language and Culture

The importance of milk in human culture is attested to by the numerous expressions embedded in our languages, for example "the milk of human kindness". In ancient Greek mythology, the goddess Hera spilled her breast milk after refusing to feed Heracles, resulting in the Milky Way.

In African and Asian developing nations, butter is traditionally made from fermented milk rather than cream. It can take several hours of churning to produce workable butter grains from fermented milk.

Holy books have also mentioned milk; the Bible contains references to the 'Land of Milk and Honey'. In the Quran, there is a request to wonder on milk as follows: 'And surely in the livestock there is a lesson for you, We give you to drink of that which is in their bellies from the midst of digested food and blood, pure milk palatable for the drinkers.'(16-The Honeybee, 66). The Ramadhan fast is traditionally broken with a glass of milk and dates.

The verb, "to milk" something is often used in the vernacular of many English-speaking countries as a synonym for extortion or, in less loaded terms, taking advantage of a situation where one has another person at a disadvantage, as in 'milking the situation'.

The word milk has had many slang meanings over time. In the early 17th century the word was used to mean semen, or vaginal secretions, or to masturbate oneself or someone else. In the 19th century, milk was used to describe a cheap alcoholic drink made from methylated spirits mixed with water. The word was also used to mean defraud, to be idle, to intercept telegrams addressed to someone else, and a weakling or 'milksop'. In the mid 1930s, the word was used in Australia meaning to siphon gas from a car.

Milk is sometimes referred to as moo juice in American English, while Cockney rhyming slang calls it Acker Bilk, Tom Silk, Lady in silk and Kilroy Silk.

The name of the Russian Molokan religion in Russian is derived from Russian "Ìîëîêîì " meaning "Milk" as they would drink milk on the Russian Orthodox days of fast.

Other Uses

Besides serving as a beverage or source of food, milk has been described as used by farmers and gardeners as an organic fungicide and foliage fertilizer. Diluted milk solutions have been demonstrated to provide an effective method of preventing powdery mildew on grape vines, while showing it is unlikely to harm the plant.

Cream and Butter

Today, milk is separated by large machines in bulk into cream and skim milk. The cream is processed to produce various consumer products, depending on its thickness, its suitability for culinary uses and consumer demand, which differs from place to place and country to country. Some cream is dried and powdered, some is condensed (by evaporation) mixed with varying amounts of sugar and canned. Most cream from New Zealand and Australian factories is made into butter. This is done by churning the cream until the fat globules coagulate and form a monolithic mass. This butter mass is washed and, sometimes, salted to improve keeping qualities. The residual buttermilk goes on to further processing. The butter is packaged (25 to 50 kg boxes) and chilled for storage and sale. At a later stage these packages are broken down into home-consumption sized packs. Butter sells for about US$3200 a tonne on the international market in 2007 (an unusual high).

Skimmed Milk

The product left after the cream is removed is called skim, or skimmed, milk. Reacting skim milk with rennet or with an acid makes casein curds from the milk solids in skim milk, with whey as a residual. To make a consumable liquid a portion of cream is returned to the skim milk to make low fat milk (semi-skimmed) for human consumption. By varying the amount of cream returned, producers can make a variety of low-fat milks to suit their local market. Other products, such as calcium, vitamin D, and flavouring, are also added to appeal to consumers.

Casein

Casein is the predominant phosphoprotein found in fresh milk. It has a very wide range of uses from being a filler for human foods, such as in ice cream, to the manufacture of products such as fabric, adhesives, and plastics. However, in the United States these assorted non-food uses have led to concerns over the import of substandard (non-food-grade) powders from other countries, such as China, that are then used to artificially bolster domestic cheese yield without the casein additive undergoing Food and Drug Administration inspection.

Cheese

Cheese is another product made from milk. Whole milk is reacted to form curds that can be compressed, processed and stored to form cheese. In countries where milk is legally allowed to be processed without pasteurisation a wide range of cheeses can be made using the bacteria naturally in the milk. In most other countries, the range of cheeses is smaller and the use of artificial cheese curing is greater. Whey is also the by-product of this process.

Cheese has historically been an important way of "storing" milk over the year, and carrying over its nutritional value between prosperous years and fallow ones. It is a food product that, with bread and beer, dates back to prehistory in Middle Eastern and European cultures, and like them is subject to innumerable variety and local specificity. Although nowhere near as big as the market for cow's milk cheese, a considerable amount of cheese is made commercially from other milks, especially goat and sheep.

Milk Powders

Milk is also processed by various drying processes into powders. Whole milk, skim milk, buttermilk, and whey products are dried into a powder form and used for human and animal consumption. The main difference between production of powders for human or for animal consumption is in the protection of the process and the product from contamination. Some people drink milk reconstituted from powdered milk, because milk is about 88% water and it is much cheaper to transport the dried product. Dried skim milk powder is worth about US$5300 a tonne (mid-2007 prices) on the international market.

Other Milk Products

Kumis is produced commercially in Central Asia. Although it is traditionally made from mare's milk, modern industrial variants may use cow's milk instead.

Transport of Milk

Historically, the milking and the processing took place in the same place: on a dairy farm. Later, cream was separated from the milk by machine, on the farm, and the cream was transported to a factory for buttermaking. The skim milk was fed to pigs. This allowed for the high cost of transport (taking the smallest volume high-value product), primitive trucks and the poor quality of roads. Only farms close to factories could afford to take whole milk, which was essential for cheese making in industrial quantities, to them. The development of refrigeration and better road transport, in the late 1950s, has meant that most farmers milk their cows and only temporarily store the milk in large refrigerated bulk tanks, whence it is later transported by truck to central processing facilities.

Milking Machines

Milking machines are used to harvest milk from cows when manual milking becomes inefficient or labour intensive. The milking unit is the portion of a milking machine for removing milk from an udder. It is made up of a claw, four teatcups, (Shells and rubber liners) long milk tube, long pulsation tube, and a pulsator. The claw is an assembly that connects the short pulse tubes and short milk tubes from the teatcups to the long pulse tube and long milk tube. (Cluster assembly) Claws are commonly made of stainless steel or plastic or both. Teatcups are composed of a rigid outer shell (stainless steel or plastic) that holds a soft inner liner or inflation. Transparent sections in the shell may allow viewing of liner collapse and milk flow. The annular space between the shell and liner is called the pulse chamber.

Milking machines work in a way that is different from hand milking or calf suckling. Continuous vacuum is applied inside the soft liner to massage milk from the teat by creating a pressure difference across the teat canal (or opening at the end of the teat). Vacuum also helps keep the machine attached to the cow. The vacuum applied to the teat causes congestion of teat tissues (accumulation of blood and other fluids). Atmospheric air is admitted into the pulsation chamber about once per second (the pulsation rate) to allow the liner to collapse around the end of teat and relieve congestion in the teat tissue. The ratio of the time that the liner is open (milking phase) and closed (rest phase) is called the pulsation ratio.

The four streams of milk from the teatcups are usually combined in the claw and transported to the milkline, or the collection bucket (usually sized to the output of one cow) in a single milk hose. Milk is then transported (manually in buckets) or with a combination of airflow and mechanical pump to a central storage vat or bulk tank. Milk is refrigerated on the farm in most countries either by passing through a heat-exchanger or in the bulk tank, or both.

In the photo above is a bucket milking system with the stainless steel bucket visible on the far side of the cow. The two rigid stainless steel teatcup shells applied to the front two quarters of the udder are visible. The top of the flexible liner is visible at the top of the shells as are the short milk tubes and short pulsation tubes extending from the bottom of the shells to the claw. The bottom of the claw is transparent to allow observation of milk flow. When milking is completed the vacuum to the milking unit is shut off and the teatcups are removed. Milking machines keep the milk enclosed and safe from external contamination. The interior 'milk contact' surfaces of the machine are kept clean by a manual or automated washing procedures implemented after milking is completed. Milk contact surfaces must comply with regulations requiring food-grade materials (typically stainless steel and special plastics and rubber compounds) and are easily cleaned. Most milking machines are powered by electricity but, in case of electrical failure, there can be an alternative means of motive power, often an internal combustion engine, for the vacuum and milk pumps. Milk cows cannot tolerate delays in scheduled milking without serious milk production reductions.

Animal Waste from Large Dairies

As measured in phosphorus, the waste output of 5,000 cows roughly equals a municipality of 70,000 people. In the U.S., dairy operations with more than 1,000 cows meet the EPA definition of a CAFO (Concentrated Animal Feeding Operation), and are subject to EPA regulations. For example, in the San Joaquin Valley of California a number of dairies have been established on a very large scale. Each dairy consists of several modern milking parlour set-ups operated as a single enterprise. Each milking parlour is surrounded by a set of 3 or 4 loafing barns housing 1,500 or 2,000 cattle. Some of the larger dairies have planned 10 or more series of loafing barns and milking parlours in this arrangement, so that the total operation may include as many as 15,000 or 20,000 cows. The milking process for these dairies is similar to a smaller dairy with a single milking parlour but repeated several times.

The size and concentration of cattle creates major environmental issues associated with manure handling and disposal, which requires substantial areas of cropland (a ratio of 5 or 6 cows to the acre, or several thousand acres for dairies of this size) for manure spreading and dispersion, or several-acre methane digesters. Air pollution from methane gas associated with manure management also is a major concern. As a result, proposals to develop dairies of this size can be controversial and provoke substantial opposition from environmentalists including the Sierra Club and local activists.

The potential impact of large dairies was demonstrated when a massive manure spill occurred on a 5,000-cow dairy in Upstate New York, contaminating a 20-mile (32 km) stretch of the Black River, and killing 375,000 fish. On Aug. 10, 2005, a manure storage lagoon collapsed releasing several million gallons of manure into the Black River. Subsequently the New York Department of Environmental Conservation mandated a settlement package of $2.2 million against the dairy.

Use of Hormones

It is possible to maintain higher milk production by injecting cows with growth hormones known as recombinant BST or rBGH, but this is controversial due to its effects on animal and possibly human health. The European Union, Japan, Australia, New Zealand and Canada have banned its use due to these concerns. However, no such prohibition exists in the US, where approximately 17.2% of dairy cows are treated in this way. The U.S. Food and Drug Administration states that no "significant difference" has been found between milk from treated and non-treated cows but based on consumer concerns several milk purchasers and resellers have elected not to purchase milk produced with rBST.

Management of the Herd

Modern dairy farmers use milking machines and sophisticated plumbing systems to harvest and store the milk from the cows, which are usually milked two or three times daily. During the warm months, in the northern hemisphere, cows may be allowed to graze in their pastures, both day and night, and are brought into the barn only to be milked. Many barns also incorporate tunnel ventilation into the architecture of the barn structure.

This ventilation system is highly efficient and involves opening both ends of the structure allowing cool air to blow through the building. Farmers with this type of structure keep cows inside during the summer months to prevent sunburn and damage to udders.

During the winter months, especially in northern climates, the cows may spend the majority of their time inside the barn, which is warmed by their collective body heat. Even in winter, the heat produced by the cattle requires the barns to be ventilated for cooling purposes. Many modern facilities, and particularly those in tropical areas, keep all animals inside at all times to facilitate herd management. Housing the cow can be either loose housed or stalls (called cow cubicles in UK). There is little research available on dimensions required for cow stalls, and much housing can be out of date, however increasingly companies are making farmers aware of the benefits, in terms of animal welfare, health and milk production.

In the southern hemisphere cows spend most of their lives outside on pasture, although they may receive supplementation during periods of low pasture availabliity.

The production of milk requires that the cow be in lactation, which is a result of the cow having given birth to a calf. The cycle of insemination, pregnancy, parturition, and lactation, followed by a "dry" period of a few weeks before calving which allows udder tissue to regenerate. Dairy operations therefore include both the production of milk and the production of calves. Bull calves are either castrated and raised as steers for beef production or veal.

Health and well-being

Common ailments affecting dairy cows include infectious disease (e.g. mastitis, endometritis and digital dermatitis), metabolic disease (e.g. milk fever and ketosis) and injuries caused by their environment (e.g. hoof and hock lesions). Lameness is commonly considered one of the most significant animal welfare issues for dairy cattle. It can be caused by a number of sources, including infections of the hoof tissue (e.g. fungal infections that cause dermatitis) and physical damage causing bruising or lesions (e.g. ulcers or hemorrhage of the hoof). While housing and management features common in modern conventional dairy farms (such as concrete barn floors, limited access to pasture and suboptimal bed-stall design) have been identified as contributing risk factors, small farms in developing countries can also demonstrate high rates.

Market

India is the largest producer of dairy products in the world. There is a great deal of variation in the pattern of dairy production worldwide. Many countries which are large producers, consume this internally, while others — in particular New Zealand — export a large percentage of their production. Internal consumption is often in the form of liquid milk, while the bulk of international trade is in processed dairy products such as milk powder. The world's largest exporter of dairy products is New Zealand, and dairy products are the largest export earner for the country. Fonterra is the fifth-largest dairy company in the world and New Zealand's largest company by turnover,

European Union

The European Union is the largest milk producer in the world, with 143.7 million tonnes in 2003. This data, encompassing the present 25 member countries, can be further broken down into the production of the original 15 member countries, with 122 million tonnes, and the new 10 mainly former Eastern European countries with 21.7 million tonnes.

United States

In the United States, the top four dairy states are, in order by total milk production; California, Wisconsin, New York, and Idaho.

Dairy farming is also an important industry in Florida, Minnesota, Ohio and Vermont. There are 65,000 dairy farms in the United States.

Pennsylvania however, is the state with the heaviest dependence on dairy farming — there it is the number one industry. Pennsylvania is home to 8,500 farms and 555,000 dairy cows. Milk produced in Pennsylvania yields about US$1.5 billion in farm revenue every year, and is sold to various states up and down the east coast. Milk prices collapsed in 2009. Senator Bernie Sanders accused Dean Foods of controlling 40% of the country's milk market. He has requested the United States Department of Justice to pursue 'an anti-trust investigation. Dean Foods says it buys 15% of the country's raw milk.

Competition

Most milk-consuming countries have a local dairy farming industry, and most producing countries maintain significant subsidies and trade barriers to protect domestic producers from foreign competition. In large countries, dairy farming tends to be geographically clustered in regions with abundant natural water supplies (both for feed crops and for cattle) and relatively inexpensive land (even under the most generous subsidy regimes, dairy farms have poor return on capital). New Zealand, the fourth largest dairy producing country, does not apply any subsidies to dairy production.

The milking of cows was traditionally a labour-intensive operation and still is in less developed countries. Small farms need several people to milk and care for only a few dozen cows, though for many farms these employees have traditionally been the children of the farm family, giving rise to the term "family farm".

Advances in technology have mostly led to the radical redefinition of "family farms" in industrialized countries such as the United States. With farms of hundreds of cows producing large volumes of milk, the larger and more efficient dairy farms are more able to weather severe changes in milk price and operate profitably, while "traditional" very small farms generally do not have the equity or cash flow to do so. The common public perception of large corporate farms supplanting smaller ones is generally a misconception, as many small family farms expand to take advantage of economies of scale, and incorporate the business to limit the legal liabilities of the owners and simplify such things as tax management.

Before large scale mechanization arrived in the 1950s, keeping a dozen milk cows for the sale of milk was profitable. Now most dairies must have more than one hundred cows being milked at a time in order to be profitable, with other cows and heifers waiting to be "freshened" to join the milking herd. In New Zealand the average herd size, depending on the region, is about 350 cows.

Herd size in the US varies between 1,200 on the West Coast and Southwest, where large farms are commonplace, to roughly 50 in the Northeast, where land-base is a significant limiting factor to herd size. The average herd size in the U.S. is about one hundred cows per farm. Currently, concerns regarding monopolies created by Dean Foods, Kraft, and other major buyers of bulk dairy products on the Chicago Mercantile Exchange have been raised, as American dairy farms have suffered extreme price depression and chaotic fluctuations while processors and retailers report record profits. Many theorize that unregulated imports of milk protein concentrate used by processors to boost cheese yield has artificially and unfairly influenced the markets in an effort to force consolidation and vertical integration in what has historically been a highly diversified industry.

Industrial Processing

Dairy plants process the raw milk they receive from farmers so as to extend its marketable life. Two main types of processes are employed: heat treatment to ensure the safety of milk for human consumption and to lengthen its shelf-life, and dehydrating dairy products such as butter, hard cheese and milk powders so that they can be stored.

Milk

Milk is an opaque white liquid produced by the mammary glands of mammals. It provides the primary source of nutrition for young mammals before they are able to digest other types of food. The early lactation milk is known as colostrum, and carries the mother's antibodies to the baby. It can reduce the risk of many diseases in the baby. The exact components of raw milk varies by species, but it contains significant amounts of saturated fat, protein and calcium as well as vitamin C. Cow's milk has a pH ranging from 6.4 to 6.8, making it slightly acidic.

Types of Consumption

There are two distinct types of milk consumption: a natural source of nutrition for all infant mammals, and a food product for humans of all ages derived from other animals.

Nutrition for Infant Mammals

In almost all mammals, milk is fed to infants through breastfeeding, either directly or by expressing the milk to be stored and consumed later. Some cultures, historically or currently, continue to use breast milk to feed their children until they are 7 years old.

In a short time, feeding infants fresh goat's milk in lieu of breast milk has become a common practice in Western culture. There are known dangers in this practice, including risk of developing electrolyte imbalances, metabolic acidosis, megaloblastic anemia and a host of allergic reactions.

Food Product for Humans

In many cultures of the world, especially the Western world, humans continue to consume milk beyond infancy, using the milk of other animals (especially cattle, goats and sheep) as a food product. For millennia, cow's milk has been processed into dairy products such as cream, butter, yogurt, kefir, ice cream, and especially the more durable and easily transportable product, cheese. Modern industrial processes produce casein, whey protein, lactose, condensed milk, powdered milk, and many other food-additive and industrial products.

Humans are an exception in the natural world for consuming milk past infancy, despite the fact that more than 75% of adult humans show some degree (some as little as 5%) of lactose intolerance, a characteristic that is more prevalent among individuals of African or Asian descent. The sugar lactose is found only in milk, forsythia flowers, and a few tropical shrubs. The enzyme needed to digest lactose, lactase, reaches its highest levels in the small intestines after birth and then begins a slow decline unless milk is consumed regularly. On the other hand, those groups that do continue to tolerate milk often have exercised great creativity in using the milk of domesticated ungulates, not only of cattle, but also sheep, goats, yaks, water buffalo, horses, and camels. The largest producer and consumer of cattle and buffalo milk in the world is India.

Bibliography

Adams, C.E. : *Mammalian Egg Transfer*, Boca Raton, FL, CRC Press, 1982.

Adams, Carol J.: *Animals and Women: Feminist Theoretical Explorations.* Durham, NC: Duke University Press, 1995.

Archana Satarkar: *Food Science and Nutrition*, ABD Pub, Delhi, 2008.

Arora, N. : *Manual of Animal Nutrition*, International Book, 2004.

Arseniev, V.A.: *Atlas of Marine Mammals*, Neptune City: T. F. H. Publishing, Inc., 1986.

Arti Sharma: *Fishes : Aid to Collection, Preservation and Identification*, Daya, Delhi, 2006.

Aruna T. Kumar : *Handbook of Animal Husbandry*, Indian Council of Agricultural Research, 2008.

Baker, Steve: *The Postmodern Animal.* London: Reaktion Books, 2000.

Balram Pani: *Textbook of Animal Chemistry*, I K International, Delhi, 2007.

Basavaraj S. Benni: *Dairy Co-operative Management and Practice*, Rawat, Delhi, 2005.

Baudrillard: *The Animals: Territory and Metamorphoses. Simulacra and Simulation.* Ann Arbor: University of Michigan Press, 1994.

Bekoff, Marc: *Strolling with Our Kin: Speaking For and Respecting Voiceless Animals.* New York: Lantern Books, 2000.

Betteridge, K.J. : *Embryo Transfer in Farm Animals,* Ottawa, Agriculture Canada, 1977.

Billingham, R.E., and W.K. Silvers. : *Transplantation of Tissues and Cells*, Wistar Inst. Press, Philadelphia, 1961.

Bourbon, Richard M.: *Understanding Animal Breeding*, Prentice-Hall, 2000.

Bower, B.: *Fossils may Clarify Mammal Evolution*, Science News, 1984.

Brock, J. : *A Natural History of Domesticated Animals*, Cambridge Univ. Pr., New York, 1999.

Bronson, F. H.: *Mammalian Reproductive Biology*, Univ. Chicago Pr., Chicago, 1990.

Brown, L.: *Cruelty to Animals: The Moral Debt.* London: MacMillan, 1988.

Brown, R. E.: *Social Odours in Animals Reproduction*, Clarendon Press, Oxford, 1985.

Bushnell, R.B. . : *Dry Cow Feeding and Management*, A Western Regional Extension Publication, 1979.

Carroll, R. L.: *Vertebrate Paleontology and Evolution*. W. H. Freeman and Co., New York, 1988.

Clark, Stephen: *The Moral Status of Animals*. Oxford: Oxford University Press, 1977.

Clutton Brock Juliet : *Horse power: a history of the horse and donkey in human societies*, National history Museum publications, London 1992.

Clymer, R. : *Nature's Healing Agents*, PA, U.S.A: Dorrance Co., 1963.

Crawford A. : *Experiments and Observations on Animal Heat*, London: Printed for J. Johnson; 1788.

Daniel, J.C. Jr. : *Methods in Mammalian Reproduction*, Orlando, FL, Academic Press, 1978.

Davis, A. : *Let's Eat Right to Keep Fit*, New York, U.S.A: Harcourt Brace Jovanovich, Inc., 1970.

Degen, A. A. : *Ecophysiology of Small Desert Mammals*, Springer, New York, 1997.

DeGrazia, David: *Animals Rights: A Very Short Introduction*. Oxford: Oxford University Press, 2002.

Devender Pratap Singh : *A Handbook of Beekeeping*, Agrobios, 2006.

Devyani Khemka: *Animal Physiology*, Dominant, Delhi, 2003.

Eisenberg, John E.: *The Animal Radiations*, The University of Chicago Press, 1981.

Ensminger, M.E. : *Dairy Cattle Science*, The Interstate Printers & Publishers, Inc., Danville, 1980.

Escobar, Roberto Calle: *Animal Breeding and Production of Camelids*, Lima, Peru, 1984.

Flowerdew, J. R. : *Animals: Their Reproductive Biology and Population Ecology*, Cambridge Univ. Pr., New York, 1987.

Gay, W.I. : *Methods of Animal Experimentation*, Academic Press, New York, 1965.

Godthelp, *Animal: Riversleigh. The Story of Animals Reproduction in Ancient Rainforests*, Reed Books, Balgowhah, 1991.

Goel, A K : *Basic Concept of Animal Chemistry*, Pearl Books, Delhi, 2008.

Gordon, G. A. : *Animals Physiology*, Harper and Row, New York, 1989.

Gray, J.: *Animal Locomotion*, Norton, New York. 1968.

Greene, H. W.: *Mode of Reproduction in Lizards and Snakes of the Gomez Farias Region*, Tamaulipas, Mexico. Copeia, 1970.

Griffin, D.R.: *Animal Minds*, University of Chicago Press. Chicago, 1992.

Grzimek, B.: *Grzimek's Animal Life Encyclopedia*, McGraw Hill, New York, 1989.

Hacker, J.B. : *Nutritional Limits to Animal Production from Pasture,* Farnham Royal: CAB, 1981.

Hagedorn, A.L.: *Animal Breeding*, Crosby Lockwood, 1950.

Harrison, R. J.: *Functional Anatomy of Marine Animals*, New York: Academic Press, 1974.

Hulbert AJ, Else PL. : *Mechanisms Underlying the Cost of Living in Animals,* Annu Rev Physiol. 2000.

Hutt, Frederick B.: *Genetics for Dog Breeders*, Freeman & Company, 1979.

Joysey, K. A. : *Development, Function and Evolution of Animal Teeth*, Academic Pr., New York, 1978.

Krieger, Maggie and Richard: *Secrets of the Andean Alpaca - The Field Guide*, Saltspring Island Llamas and Alpacas, 1994.

Lance, J.W. : *Migraine and Other Headaches,* New York, U.S.A: Scribner, 1986.

Lata Bhattacharya: *Animal Biochemistry*, Discovery, Delhi, 2010.

Lyster, S. : *Animals and Their Moral Standing.* London : Routledge, 1997.

Marshall, R.B. : *Breeding Farm Animals*, Asiatic Pub, Delhi, 2006.

Martin, A.M. : *Fisheries Processing : Biotechnological Applications*, Chapman and Hall, Delhi, 2009.

Mathialagan, P : *Textbook of Animal Husbandry and Livestock Extension,* International Book Distributing Co, Delhi, 2005.

Matthews, L. H.: *The Life of Animals Reproduction*, London, Weidenfield and Nicholson, 1969.

Mindell, E. : *Mindell's Vitamin Bible,* New York, U.S.A: Warner Books, 1980.

Montgomery, G. G.: *The Early Placental Mammal Radiation Using Bayesian Phylogenetics*, Science, December 2001.

Muybridge, E. : *Muybridge's Complete Human and Animal Locomotion*, Dover Publ., New York, 1979.

Nyholt, D.H. : *The Vitamin & Herb Guide,* Alberta, Canada: Global Health Ltd., 1992.

Rathnakumar, K : *Fish Processing Technology and Product Development*, Narendra Pub, Delhi, 2008.

Raymond, F., Redman, P., & Waltham, R. : *Forage conservation and Feeding.* Ipswich: Farming Press, 1986.

Renaville, R and A Burny : *Biotechnology in Animal Husbandry*, Springer Pub, 2008.

Rhykerd Charles L. : *The Cycles of Plant and Animal Nutrition*, Scientific American Books, San Francisco 1976.

Robinson, Roy: *Genetics for Dog Breeders*, Pergamon Press, 1990.

Safley, Michael: *Synthesis of a Miracle*, Northwest Alpacas, 2002.

Schiller, A. L. : *Anatomy of the Guinea Pig*, Harvard Univ. Pr., Cambridge, 1975.

Seidel, S.M.: *New Technologies in Animal Breeding*, Orlando, FL, Academic Press, 1981.

Shagufta Jamal and H P S Arya : *Participatory Rural Appraisal in Agriculture and Animal Husbandry : A Training Manual*, Concept, 2004.

Short, R. V.: *Reproduction in Mammals*, Cambridge, Cambridge University Press, 1972.

Shukla, M K : *Brain Teasers : Multiple Choice Questions on Animal Husbandry and Veterinary Sciences*, International Book Dist, Delhi, 2007.

Singh,. G : *Chemistry of Amino-Acids and Proteins*, Discovery, Delhi, 2007.

Stuart Patton : *Principles of Dairy Chemistry*, Huntington, N.Y.: Krieger, 1976.

Thornhill, Nancy W.: *The Natural History of Inbreeding and Outbreeding*, Chicago Press, 1993.

Verma, S.R. : *Nature: Fish Genetics and Biodiversity Conservation*, Conservators, Delhi, 1998.

White, M.J.D.: *Animal Cytology and Evolution*, Cambridge, Cambridge Univ. Press, 1954.

Yablokov, A.: *Variability of Mammals: Moscow*, USSR, Nauka Publishers, 1966.

Yadav, Manju: *Mammalian Development*, Discovery Publishing House, Delhi, 2008.

Index

S

T

W